T0250206

High Solids Alkyd Resins

High Solids Alkyd Resins

Krister Holmberg
Berol Kemi AB
Stenungsund, Sweden

Taylor & Francis
Taylor & Francis Group

LONDON AND NEW YORK

Published by Taylor & Francis
2 Park Square, Milton Park, Abingdon, Oxon, OX14 4RN
270 Madison Ave, New York NY 10016

Transferred to Digital Printing 2010

Library of Congress Cataloging-in-Publication Data

Holmberg, Krister
    High solids alkyd resins.

    Includes index.
    1. High solids coatings. 2. Alkyd resins. I. Title.
TP945.H65 1987   667'.9   87-20125
ISBN 0-8247-7778-6

Publisher's Note
The publisher has gone to great lengths to ensure the quality of this reprint
but points out that some imperfections in the original may be apparent.

Preface

High solids coatings are not new. Vegetable oils, especially linseed oil, have a long tradition both as high solids binders and as reactive solvents for other binders. In earlier days, however, a low solvent emission was not an important matter for the coatings industry. As new vehicles were developed that performed better but had a higher solvent demand, the industry rapidly switched to these.

During the last 10 to 20 years great pressure has been placed on the paint formulator to reduce the emission of organic solvents. "High solids" is an expression that has become known to everyone in the field. The goal to reduce the amount of volatile components in formulations is common with almost all types of paints and lacquers, air-drying as well as enamels. The same trend can be found almost everywhere where organic solvents are being used.

For alkyd-based paints there are two main alternatives to reduced levels of volatile organic solvents: high solids and water-reducible systems. This volume deals with high solids. The predominant aim of this monograph is to cover the chemistry, as opposed to the technology, of high solids compositions. This means that emphasis is put on the binder component and, to a lesser extent, on the solvent. Formulations of high solids surface coatings are only dealt with in passing. Appropriate formulations are usually available from the raw material suppliers.

Factors controlling the viscosity and, consequently, the solids contents of alkyd resins are discussed in relative detail. Different approaches mentioned in the literature to preparing high solids alkyds are described. Since the solids content of the binder is intimately linked to the choice of solvent, Chapter 4 is devoted to this subject.

Alkyd resins for stoving are generally combined with amino resins, usually melamine- or urea-formaldehyde condensates. Melamine resins are preferred in high solids formulations. Systems of this type are discussed in Chapter 5.

The topic of high solids alkyds would not be complete unless a chapter dealing with reactive solvents for these binders were included. The emphasis in Chapter 6 is on newer developments in the area. The traditional system of this type, i.e., styrene-unsaturated polyester, is also described.

Krister Holmberg

Contents

1

Introduction

1.1 THE MEANING OF HIGH SOLIDS

High solids coatings is only one of several approaches to solvent-
less or solvent-free paints. The move toward a higher nonvolatile
content for paints and lacquers can be said to have been initiated
in 1966 by the well-known "Rule 66" of California, which regu-
lates the emission levels of photochemically reactive solvents in
paint formulations. However, it is only during the last decade that
paints with a considerably higher solids content have entered the
market.

Strictly speaking, the term "high solids" refers to coatings
with more than 80% nonvolatiles by volume. In practice, paints
with 70% and even 60% volume solids are usually included in the
high solids category. However, since the higher level is necessary
for exemption under air pollution legislation, this is the goal that
the industry is striving for.

Because the solvent is the component of a paint formulation with the lowest density, the 80% volume figure corresponds
to 85-88% weight solids. Thus, to fulfill the strict demand of a
high solids formulation, only 12-15 g of solvent (the exact amount
depending mainly on the type and amount of pigment used) can
be employed. Achieving this level without affecting drying and
film properties in a negative way places extreme demands on the
binder, as well as on the solvent.

Throughout the book, the term "high solids" is used in a
free sense and is not restricted to formulations having 80% or
more nonvolatiles by volume.

1.2 DRIVING FORCES

The main driving forces for the reduction of solvents in paints and lacquers are

air pollution control
control of indoor emission of solvents
energy savings
materials savings
time savings

The last three points refer to the evaporation of solvent during the drying and curing operations. Even if a considerable amount of resources were wasted in these processes, the pollution aspects, exterior and interior, are by far the most important driving forces for the development of high solids paints. In the following, air pollution regulations will be briefly discussed. The reader is referred to Ref. 1 for a more comprehensive review of the air pollution regulations in the United States and in various European countries.

The objective of Rule 66 of California was to control the emission level of "photochemically reactive" solvents. Major emphasis was given to aromatic solvents and to some ketones. As a consequence, the concept of so-called exempt solvents was born and rapidly used in many types of paints.

It gradually became clear, however, that a division of solvents into these two categories is not adequate. Harmless organic compounds can, for instance, become irritants by reactions in the atmosphere induced by ultraviolet light.

In accordance with this view the U.S. Environmental Protection Agency (EPA) has enforced the concept of Volatile Organic Compounds (VOC) as its measure of emission control. All components in a paint—not only solvents—are defined as volatile if the vapor pressure exceeds 13.3 Pa (0.1 mm Hg) at 25°C. (There are still exceptions, discussed later.)

The amount of solvent emitted during the coating process depends not only on the solids content of the paint but also on the efficiency of the application process. In the EPA guidelines, as well as in the Clean Air Act of 1977, *total* emission of volatile compounds is discussed, including application losses.

The VOC requirements of paints also vary from industry to industry and depend on whether the coatings are stoved or air dried at temperatures below 90°C. For baking enamels with stoving temperatures above 90°C, lower VOC values are required.

The VOC of a paint is measured on the paint as it is being used, according to the expression

$$\text{VOC} = \rho \, \frac{100 - \text{NV}_s}{100}$$

where ρ = density of the paint in kg/m^3 (or g/l)
   NV_s = % nonvolatile by weight

The VOC value is then obtained in kg/m^3 (or g/l).

Also in the EPA guidelines two solvents, dichloromethane and 1,1,1-trichloroethane, have been exempted from control. These two chlorinated solvents are considered to have negligible photochemical reactivity in the troposphere and no significant impact in the stratosphere. This has led to an increased use of these two solvents in high solids and radiation curing systems.

In Europe many countries have differentiated the solvents according to their harmfulness. West Germany is the forerunner in this respect with its Federal Emission Law of 1974. The volatile emissions are divided into three classes, with Class I compounds having extremely low emission limits. Most of the commonly used paint solvents belong to Classes II or III. The limits for each class have gradually been reduced since the law was issued and some solvents have been reclassified. Most West European countries have regulations or recommendations which basically follow either the United States or West Germany. The situation is very complex, however, and many countries have regional air pollution regulations.

1.3 EMERGING TECHNOLOGIES

The high solids concept is not the only development line being pursued in order to reduce the amount of organic solvent in paints and lacquers. A number of new technologies, all derived from the demand to reduce or eliminate the organic solvent from the coating system, have become established on the market today. The most important of these, apart from high solids coatings, are powder coatings, waterborne coatings, radiation curable coatings, and reactive diluent systems. Each one of these will be discussed briefly later.

1.3.1 Powder Coatings

Powder coatings may be regarded as the ultimate form of high solids systems. They have been on the market for about two decades by now, but their breakthrough, which has been predicted is yet to be seen. Powder coatings are today being used in applications such as household products, electrical devices, and in the automotive industry. Besides the very low emission of volatile compounds, powder coatings have a number of advantages, such as high application efficiency, low energy consumption, and low labor and clean-up costs.

Powder coatings may be based on either thermoplastic or thermosetting polymers. The former are mainly applied by a fluid-bed process, and no chemical curing process takes place after application. The thermosetting polymers are normally applied by electrostatic spray processes, and the material undergoes polymerization on heating after application.

Representative examples of thermoplastic and thermosetting polymers are as follows:

Thermoplastic	Thermosetting
Polyesters	Polyesters
Polyamides	Polyacrylates
Polyolefins	Epoxy resins
Cellulose esters	

High quality finishes are obtained with thermosetting two-component systems. The curing conditions are typically 15 min at 140-160°C, and application by modern high voltage powder guns can give relatively thick, smooth films even on irregular surfaces. Oversprayed powder can be recovered to a large extent and re-used. The waste in powder coating operations is, therefore, very small.

The main curing reactions taking place with thermosetting powder coatings are given in Figure 1.1. In the majority of powder coating compositions epoxy resins are involved either as the main component or as the cross-linker.

References 2-5 give deeper descriptions of the area.

1.3.2 Waterborne Coatings

Waterborne coatings may be subdivided into two major groups, aqueous dispersions having a particle size in the range of 0.1-3

Main Polymer   Curing Reaction

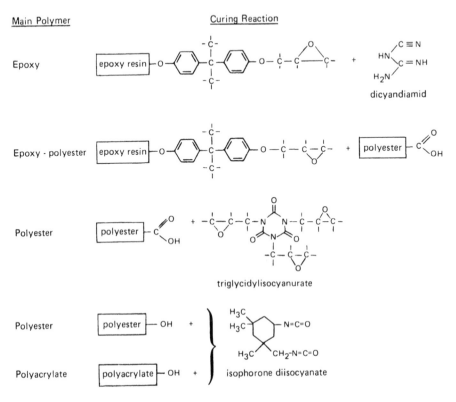

Figure 1.1 Powder coatings systems.

μm and water reducible coatings with a particle size below 0.01
μm. Coating systems of intermediate particle size (0.01-0.1 μm)
have also been described, for instance in the context of microemulsions of alkyds, but they seem not to have reached the market on a larger scale yet.

Aqueous dispersions or latexes are composed of polymer
particles dispersed in water and stabilized with surfactants. Both
thermoplastic and thermosetting systems of this type are so well
established on the market that they, as a group, can hardly be
included within the concept of emerging technologies. However, new advanced fields are being developed within the frame-

work of aqueous dispersions. Polyurethane dispersions [5] are probably the best example of this.

Water-reducible coatings are usually made by employing polymers possessing either carboxylic or amino groups that are partially or fully neutralized with base or acid, respectively. Carboxyl-containing alkyds and acrylics neutralized with amines are the prime examples of this class of coatings. A water-miscible solvent (coupling solvent) must usually be added as well, and, in practice, the VOC may be rather high.

Water-reducible resins have an established position in electrodeposition coating compositions. These can be either anionic, e.g., polybutadiene adducts containing 5-25% maleic anhydride and often modified with phenolic resin, or cationic, e.g., amino functional epoxy resins.

Introduction of polyglycol ether chains into the alkyd or polyacrylate resin molecule is an alternate way of introducing a certain water solubility. This has been suggested as a way to eliminate the use of volatile amines and coupling solvents [6]. Water sensitivity of the cured film and poor adhesion to many substrates are, however, severe drawbacks of this approach.

Water-reducible coatings are reviewed in Ref. 4.

1.3.3 Radiation Curable Coatings

Radiation-curable coatings comprise ultraviolet (UV), electron beam (EB), intrared (IR), and microwave cured systems. Only UV and EB curing seem to be of commercial importance for paints and lacquers today. It should be noted, however, that IR curing is being used on a large scale in the related paper-coating process.

The energy input and the type of curing mechanism for the various processes are shown in Table 1.1.

Table 1.1 Energy Input and Curing Mechanisms for Various Radiation Curing Processes

Source	Energy (ev)	Curing Mechanism
Electron beam	10^5	Ionization excitation
Ultraviolet	5	Electronic excitation
Microwave	10^{-3}	Thermal
Infrared	10^{-1}	Thermal

UV curing systems is by far the most common of the radiation processes. The main components of such a system are the following:

1. A low-molecular resin containing olefinic bonds.
2. A reactive solvent, also containing unsaturated groups.
3. A photoinitiator.

Originally, unsaturated polyesters were used as the polymer in UV curing systems. However, these systems have a relatively low curing rate and are of limited use today. Instead, acrylate-based pre-polymers are widely used. These may either be polyacrylate resins or some other polymer, e.g., alkyd, epoxy, or polyurethane, modified with acrylate groups. Acrylated polyesters, prepared by esterification of free hydroxyl groups of the polymer with acrylic acid, have gained widespread use.

Acrylic esters are normally used for the reactive solvent. Mixtures of monofunctional and di- or polyfunctional monomers are frequently employed. Acrylated polyols, e.g., trimethylolpropane triacrylate and pentaerythritol triacrylate, have been found to be particularly effective in combination with a number of polymers [7]. Styrene is used as the main monomer in combination with nonmodified unsaturated polyesters.

The role of the photoinitiator is to generate active species through light absorption. These species subsequently disintegrate to radicals and ions, which bring about the polymerization in the system. Acetophenones, benzophenones, benzil derivatives, and benzoin ethers are the most commonly used initiators.

UV curing systems have an established position in wood coatings. They have reached a commercial position in a number of other areas as well, but a major breakthrough has not yet occurred. Problems such as oxygen inhibition, light absorption, and scattering by pigments and the toxicity of the monomers seem to be severe drawbacks for the technique in the future.

References 4 and 8 are review articles on radiation curable coatings.

1.3.4 Reactive Diluent Systems

UV curing systems may be regarded as a good example of reactive diluent systems. However, by convention only thermally cured systems are included in this category, and systems cured by irradiation are normally dealt with separately.

Reactive diluent systems are closely related to high solids systems, and the subject is treated in Chapter 6. The reactive diluent concept should probably not be regarded as a development line of its own, but could be viewed as an integrated part of the development of high solids systems. A small amount of reactive solvent as part of the total volume of solvent is a way of further increasing the nonvolatile content of a high solids formulation.

For high solids alkyds the concept can be used both for air-drying and stoving coatings. In the former case the diluent must contain a group capable of participating in the autoxidation process, e.g., an allyl ether function. For stoving alkyds used in

combination with amino resins the reactive diluent may simply be
a diol or a triol, which would be incorporated into the network
by the same type of curing reactions as the alkyd resin. Care must
be taken, however, so that the polyol-diluent is not lost by evaporation during the curing step [9].

1.3.5 High Solids Coatings

High solids systems have the advantage over the other new technologies discussed in this section that the switch over from conventional systems is relatively simple. The technology is familiar
and a minimum equipment change is required for its adoption.

Initial high solids efforts were concentrated on saturated
polyester resins to be used in industrial bake finishes, mainly as
an alternative to short oil coconut or tall oil alkyds [9]. Later,
the development has continued to oil-modified polyesters, both
baking and air-drying alkyds.

Development work has also been carried out on high solids
resins other than alkyds. Thermosetting acrylics cured with melamine resins have been designed to give a high solids content [10].
High solids coatings based on epoxy resins and polyurethanes have
also been reported [11,12]. The vast majority of work on high
solids coatings has been performed on alkyds, however.

Review articles on high solids systems are found in Refs.
13-15.

1.3.6 Comparison Between the New Technologies

Naturally, each of the development routes briefly described above
has its merits and its disadvantages. Table 1.2 gives a rough comparison between the various technologies.

Table 1.2 Comparison Between Newer Technologies in the Coating Field[a]

Type		Conventional solvent-thinned	Solvent replaced	NAD	High solids liquids High solids	Solventless	Waterborne	Powder
Air pollution	Photochemical active	−	+	+	+	+++	++	+++
	Total amount	−	−	+	+	+++	++	+++
Investment needs	Fume elimination	−	−	−	++	+	−	+
	Wastewater treatment	−	−	−	−	+	−	+
Energy saving	Material	−	−	+	+	+	−	−
	Energy	−	−	+	+	+	−	−
	Labor		−	+	+	+		
Painting equipment	Conventional equipment	++	++	++	+	−	−	−
	Investment				−	−	−	−
Safety	Fire	−	−	−	∓	∓	+	∓
	Toxicity	−	−	−	∓	∓	+	+
Processability		++	++	++	+	+	−	−
Film appearance		++	++	++	+	+	−	−

a+ and − mean relative advantage and limitation.
Source: Ref. 14.

1.4   RHEOLOGY OF HIGH SOLIDS COATINGS

Rheological control is a fundamental importance to high solids systems. Sagging has, for instance, been recognized as a larger problem with high solids than with conventional paints. As discussed in Chapter 4, the choice of solvent becomes more and more critical as the nonvolatile content of the formulation is raised.

Furthermore, in order to increase the solids content, viscosities of high solids coatings are usually higher than of conventional coatings. Unless the rheological behavior is under control, this will lead to a poorer atomization during the spray process.

Rheology of paints in general has been treated at length by Patton [16], and no attempt will be made here to cover the subject. However, one point of particular importance to high solids coatings will be mentioned—the use of Casson plots.

Viscosity of a fluid has been related to shear rate in a two-parameter equation known as the Casson equation:

$$\eta^{1/2} = \eta_\infty^{1/2} + \tau^{1/2} \cdot \gamma^{-1/2} \tag{1.1}$$

where η = viscosity measured
η_∞ = infinite shear viscosity
τ = yield value
γ = shear rate

A Casson plot is obtained by applying Eq. (1.1) to a system, i.e., by plotting the square root of viscosity as a function of the reciprocal of the square root of the shear rate. This is illustrated in Figure 1.2 for fluids with various rheological behavior [15]. In a Casson plot a Newtonian fluid gives a straight, horizontal line, whereas a pseudoplastic material produces a sloped line. A thixotropic fluid gives two lines with a common intercept when measured before and after shearing. The sheared material has a lower slope.

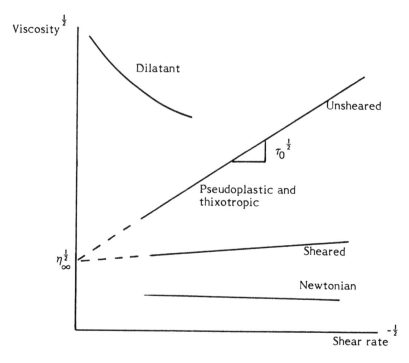

Figure 1.2 Casson plots of various fluids.

The Casson plot is extremely useful since it conveys information about two important application parameters: resistance to sag and viscosity at high shear rate.

The yield value, τ, is the minimum shear stress that must be exceeded before any flow can be observed. Below the yield value the coating acts as an elastic solid, showing deformation without flow. The yield value is also related to sag resistance; the higher the yield value, i.e., the larger the slope of a curve in the Casson plot, the more sag resistance.

The infinite shear viscosity, η_∞, is determined from the Casson plot by extrapolation. It gives an approximate value of the viscosity at very high shear rates. As can be seen from Figure 1.3,

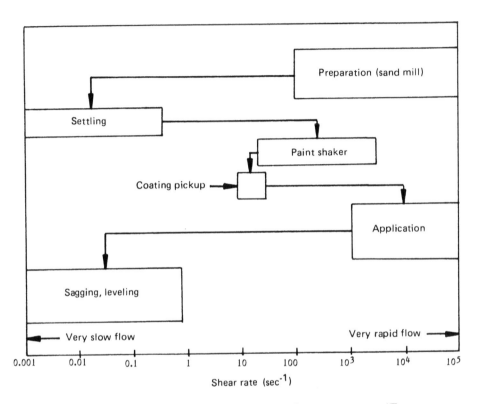

Figure 1.3 Shear rate regions of various paint processes. (From Ref. 16.)

where shear rate ranges for various coating operations are summarized, very high shear rates are present in most application procedures. In theory, η_∞ represents the viscosity at shear rates where all viscosity contributions from pigment flocculation, colloidal aggregation, etc., are eliminated. In practice, such contributions to the viscosity are considered to be unimportant at shear rates above 10,000 sec^{-1}.

In a study on high solids paints, two coatings were formulated with the same nonvolatile content [17]. Both formulations were based on the same high solids polyester-melamine resin sys-

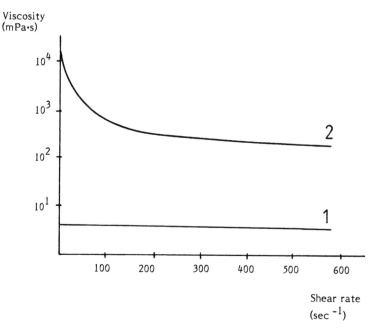

Figure 1.4 Viscosity shear profile of two high solids polyester coatings.

tem and they differed only in pigmentation. As can be seen from Figure 1.4, one (1) is an almost Newtonian fluid while the other (2) is clearly pseudoplastic. At normal shear rates the difference in viscosity between the two formulations is considerable.

A Casson plot using the viscosity vs. shear rate data of Figure 1.4 is shown in Figure 1.5. The infinite shear viscosities (η_∞ values) of paints 1 and 2 are found to be 53 mPa · s and 64 mPa · s, respectively. These results indicate that both systems could be properly atomized by a spray gun properly optimized for the respective system.

In addition, the Casson plot predicts that formulation 2 would be much more resistant to sagging than formulation 1.

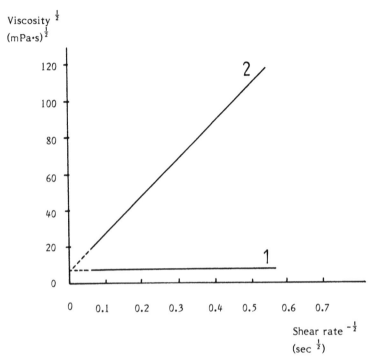

Figure 1.5 Casson plots of the high solids polyester coatings of Figure 1.4.

Probably, a flow additive would be needed to enable paint 1 to be applied in thick films.

In general, high shear viscosities are governed mainly by the binder, pigment, and solvent, whereas viscosities at low shear rates are controlled to a large extent by addition of minor amounts of rheological flow additives (gellants or thixotropic agents) [16].

Since high solids coatings are frequently formulated at higher application viscosities, conventional air spray equipment may be insufficient to effect proper atomization. Airless spray and electrostatic airless spray are more effective in spraying the higher viscosity paints. High speed electrostatic disc and bell equipment is particularly relevant to high solids paints since it takes advantage of the marked pseudoplasticity of many high solids coatings [17]. Higher voltages also facilitate atomization and deposition efficiency. The use of paint heaters is a means of reducing the viscosity to a sprayable range.

1.5 DEFINITIONS

1.5.1 Alkyds

The term "alkyd" was coined by Kienle and Ferguson and is derived from "al" of alcohol and "cid" of acid; "cid" was later changed to "kyd" [18]. Alkyd resins in a broad sense refer to polyesters. By convention, however, polyesters with unsaturation in the backbone are not referred to as alkyds but are termed "unsaturated polyesters."

There is a great deal of confusion regarding the terminology. The term alkyd is sometimes used in a restrictive manner to describe only fatty acid modified polyesters, and the nonmodified resins are then called saturated polyesters. Terms like oil-free alkyds and oil-modified polyesters can also be found in the literature.

In this book the term alkyd is used in the broader sense, comprising both oil-modified and oil-free saturated polyesters.

1.5.2 Oil Length of Alkyds

Depending on the weight % fatty acid in the resin, alkyds are referred to as short oil (less than 45%), medium oil (45-55%), or long oil (greater than 55%). The type of fatty acid also governs the properties of the alkyds, and the resins are divided into drying, semi-drying, and nondrying, depending on the degree of unsaturation in the fatty acid residues (iodine number of >140, 125-140, and <125, respectively).

The oil length and the type of oil are indicative of the properties of the alkyd resin, as shown in Table 1.3.

1.5.3 Acid Value and Hydroxyl Number

The acid value is defined as mg of potassium hydroxide required to neutralize 1 g of resin. For alkyd resins 0.1 M KOH in ethanol is normally employed.

The hydroxyl number (sometimes called hydroxyl value) is the mg of potassium hydroxide equivalent to the amount of acyl groups reacted in the acylation of 1 g of resin. A known amount of acylating reagent (often acetic anhydride or phthalic anhydride in pyridine) is added to the resin sample and the hydroxyl number is obtained by back-titration with alkali.

1.5.4 Molecular Weight and Molecular Weight Distribution

The molecular weight of a polymer is highly dependent on the measuring method used. Number average molecular weight, $\overline{M_n}$, is obtained with methods that rely on the number of molecules, whereas determinations based on the weight of the individual molecules give weight average molecular weight, $\overline{M_w}$. Unless all polymer molecules are the same size, $\overline{M_n}$ is smaller than $\overline{M_w}$. The situation is illustrated by the following mixture of an equal number of polymer molecules of molecular weights 10,000 and 30,000.

$$\overline{M_n} = 10,000 \cdot \tfrac{1}{2} + 30,000 \cdot \tfrac{1}{2} = 20,000$$

$$\overline{M_w} = \frac{10,000^2}{40,000} + \frac{30,000^2}{40,000} = 25,000$$

Table 1.3 Effect of Oil Length and Type of Oil on the Properties and Uses of Alkyds

Oil type	Oil length (%)	Typical oil	Properties
Oxidizing type	60 or more	Linseed, safflower, soya bean, tall oil fatty acids, wood oil in blends with other oils, dehydrated castor oil.	Soluble in aliphatic solvents. Compatible with oils and medium oil length alkyds, good drying characteristics. Films are flexible with reasonable gloss and durability.
Oxidizing type	45–55	Linseed, safflower, soya bean, tall oil fatty acids, wood oil in blends with other oils.	Soluble in aliphatic or aliphatic aromatic solvent mixtures. Good drying characteristics, durability, and gloss.
Oxidizing type	45 or less	Linseed, safflower, soya bean, tall oil fatty acids, wood oil in blends with other oils and dehydrated castor oil.	Soluble in aromatic hydrocarbons. Low tolerance for aliphatic solvents. Usually cured at elevated temperatures either by heating with manganese driers or with urea- or melamine-formaldehyde resins.

| Nonoxidizing type | 40-60 | Coconut oil, castor oil, hydrogenated castor oil. | Soluble in aliphatic-aromatic solvent blends. Usually used as a plasticizer for thermoplastic polymers such as nitrocellulose. |
| Nonoxidizing type | 40 or less | Coconut oil, castor oil, hydrogenated castor oil. | Soluble in aromatic solvents. Used as a reactive plasticizer which chemically combines with other resin entities, e.g., melamine formaldehyde resin. |

Source: Ref. 19.

$\overline{M_n}$ may be determined by end group analysis or by osmotic pressure. $\overline{M_w}$ is often measured by light scattering or by ultra-centrifugation. Acid value and hydroxyl number determinations, which are widely used for alkyd resins, are typical end group analyses and, thus, give the number average molecular weight.

The molecular weight distribution of a polymer is defined as the ratio of weight average to number average molecular weight, i.e., $\overline{M_w}/\overline{M_n}$. Gel permeation chromatography (GPC) is the normal measuring technique. It should be kept in mind, however, that unless a reliable calibration standard is used, GPC does not given an absolute measure of the molecular weight of each fraction.

Among the many methods developed for the molecular weight determination of polymers, the solution viscosity method is the most simple and widely used. This procedure measures the intrinsic viscosity, $[\eta]$, of a polymer in a given solvent at a constant temperature. The viscosity average molecular weight, $\overline{M_v}$, is related to $[\eta]$ by the Mark-Houwink equation:

$$[\eta] = K \cdot (\overline{M_v})^a$$

where K and a are empirical constants which depend solely upon the polymer-solvent combination.

The evaluation of the Mark-Houwink constants K and a for a particular polymer-solvent system normally requires the preparation of a series of samples having a wide range of molecular weight and their subsequent characterization, e.g., by light scattering and dilute-solution viscosity. Various GPC techniques are also used for this purpose [20,21].

The value of $\overline{M_v}$ usually lies between $\overline{M_n}$ and $\overline{M_w}$ but approaches more closely the latter. For the sake of simplicity, $\overline{M_v}$ is sometimes replaced by $\overline{M_w}$ in the Mark-Houwink equation.

A discussion of the value of the constant a for high solids systems is given in Section 2.1.

REFERENCES

1. P. G. de Lange, *Proc. XVII FATIPEC Congr.*, *3*:13 (1984).

2. G. E. Cole, Jr., *Mod. Paint Coat.*, *72*: 47 (March 1982).

3. R. C. Harrington, in *Encycl. Polym. Sci. Technol.* (H. F. Mark and N. M. Bikales, eds.), Suppl. Vol. 1, Wiley, 1976, p. 544.

4. S. Paul, *Surface Coatings Science and Technology*, Wiley, 1985, p. 561, 601, 658.

5. D. Dieterich, *Progr. Org. Coat.*, *9*:281 (1981).

6. W. J. Blank, *J. Coat. Technol.*, *49*:46 (1977).

7. C. B. Rybny, C. A. Defazio, J. K. Shahidi, J. C. Trebbellas, and J. A. Vona, *J. Coat. Technol.*, *46*:60 (1974).

8. G. W. Gruber, in *Applied Polymer Science* (J. K. Craver and R. W. Tess, eds.), ACS, ORPL, 1975, p. 304.

9. R. N. Price, *Amer. Paint Coat. J.*, *66*:50 (June 21, 1982).

10. D. I. Lunde, L. A. Wetzel, and W. J. Freund, *Mod. Paint Coat.*, *65*:23 (March 1975).

11. M. S. Chattha and H. van Oene, *Ind. Eng. Chem. Prod. Res. Dev.*, *21*:437 (1982).

12. M. Bock and W. Uerdingen, *ORPL*, *43*:52 (1980).

13. L. W. Hill and Z. W. Wicks, Jr., *Progr. Org. Coat.*, *10*:55 (1982).

14. M. Takahashi, *Polym. Plast. Technol. Eng.*, *15*:1 (1980).

15. C. K. Schoff, *Progr. Org. Coat.*, *4*:189 (1976).

16. T. C. Patton, *Paint Flow and Pigment Dispersion*, Wiley, New York, 1979.

17. R. E. Wolf, *Adv. Org. Coat. Sci. Technol. Ser.*, *4*:383 (1982).

18. R. H. Kienle and C. S. Ferguson, *Ind. Eng. Chem.*, *21*:349 (1929).

19. D. H. Solomon, *The Chemistry of Organic Film Formers*, Krieger, New York, 1977, p. 91.

20. A. R. Weiss and E. Cohn-Ginsberg, *J. Polym. Sci. B*, 7:349 (1969).

21. H. Kh. Mahabadi, *J. Appl. Polym. Sci.*, *30*:1535 (1985).

2
Factors Controlling the Solids Content of Alkyds

2.1 MOLECULAR WEIGHT

In the low molecular weight region the viscosity of a polymer in solution is more strongly influenced by its molecular weight than by any other single factor. As a consequence, the solids content of an alkyd at a given viscosity to a large extent is governed by the mean molecular weight. High solids coatings can simply be achieved by processing the alkyd to a low enough molecular weight. This approach alone is not sufficient, however, since it would lead to too much quality impairment.

As described in Section 1.5.4, the weight average molecular weight, $\overline{M_w}$, (or, more precisely, the viscosity average molecular weight, $\overline{M_v}$) is related to the intrinsic viscosity, $[\eta]$, by the Mark-Houwink equation:

$$[\eta] = K \cdot (\overline{M_w})^a \qquad (2.1)$$

where K and a are empirical constants which depend solely upon the polymer-solvent combination.

For polymer melts having M_w above a certain critical value, M_c, corresponding to an entanglement contribution to viscosity, the value of a is around 3.4. When $\overline{M_w} < M_c$, which is the situation often encountered with high solids alkyd resins (M_c is normally between 1,000 and 2,000), a is much lower, usually between 1 and 2. For concentrated solutions of polymers having $\overline{M_w} < M_c$, values of the factor a seem to vary between 0.5 and 0.8, depending on the solvating power of the solvent. It has been stated that the dependence of viscosity on $\overline{M_w}$ in high solids systems will vary with the concentration from approximatly a = 0.5 in the region of 70% solids content to a = 1-2 as 100% solids content is approached [1].

In dilute solutions a relationship between $\overline{M_w}$, $[\eta]$, and solids content, c, has been found (K is a constant):

$$\log[\eta] = K \cdot \sqrt{\overline{M_w}} \cdot c \qquad (2.2)$$

According to this equation, it is necessary to decrease $\overline{M_w}$ fourfold in order to double the solids content at the same viscosity. This is schematically illustrated in Figure 2.1. The slope of the curve varies with the choice of solvent.

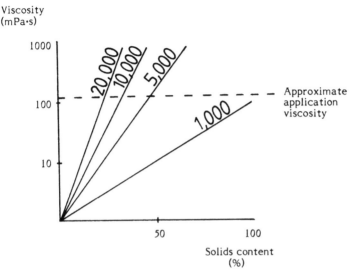

Figure 2.1 Relationship between viscosity and solids content for alkyds with varying molecular weights.

However, for highly concentrated polymer solutions Eq. (2.2) is not accurate. The discrepancy between theory and practice is illustrated by the viscosity-solids content curves of Figure 2.2a, which deviate considerably from the theoretical expectations [2]. The same data have also been plotted as isoviscosity molecular weight-percent solids curves (Fig. 2.2b) and as viscosity vs. molecular weight (Fig. 2.2c) [3]. In the latter diagram the factor a of Eq. (2.1) ranged from 1.7 to 3.3 over the three levels of solids content considered. (Note that three out of the four resins included in the study had $\overline{M_w} > \overline{M_c}$.) This illustrates the very strong dependence of molecular weight on viscosity at high solids content.

Only polymers in true solution exhibit this relationship. The viscosity of polymers in the form of emulsions or dispersions, e.g., a latex, is practically independent of the molecular

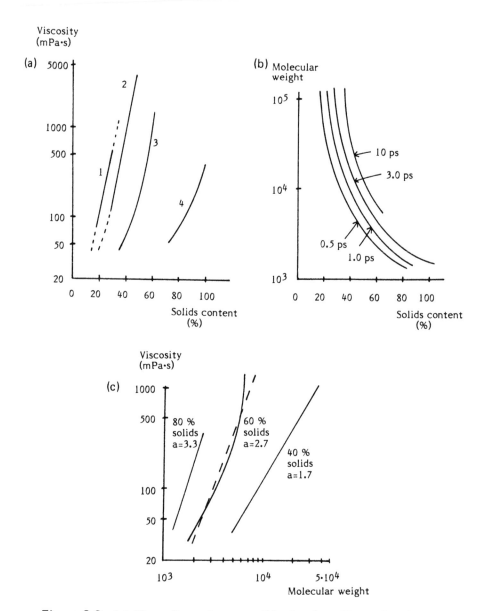

Figure 2.2 (a) Viscosity-polymer solids, isoviscosity molecular weight-polymer solids (b) and (c) viscosity-molecular weight relationships for four different polymers. 1 : Acrylic, M_w 100,000; 2 : Acrylic, M_w 25,000; 3 : Polyester, M_w 6,000; 4 : Polyester, M_w 1,500. (From Refs. 2 and 3.)

weight. The viscosity of such systems is largely governed by the viscosity of the continuous phase. Even microemulsions of poly-mers, which from a physical point of view are more related to true solutions than to emulsions, show a very weak viscosity-molecular weight relationship.

Emulsions and microemulsions of alkyds fall beyond the scope of this book. The phenomenon as such, i.e., the fact that small particles and other nonsolutes of high molecular weight may be present without greatly affecting the viscosity, is, however, of potential interest for the preparation of high solids alkyd resins. The topic is discussed further in Section 3.6.

The molecular weight of an alkyd is determined by the ratio, r_0, of carboxyl groups to hydroxyl groups. It can be shown (see p. 51) that for an ideal reaction between bifunctional reagents the following relationship exists:

$$\overline{DP} = \frac{r_0 + 1}{1 - r_0}$$

$\overline{DP}$ is the average degree of polymerization (see Section 3.1.2 for further discussion) and is defined as:

$$\overline{DP} = \frac{\text{monomers in the system}}{\text{molecules in the system}}$$

For linear polyesters in the low molecular range there also exists a linear relationship between $\log \eta$ and r_0 [4-6]:

$$\log \eta = b + c \cdot r_0$$

Resins of different composition give different values for the constants b and c, as is shown in Figure 2.3. By determining b and c for different sets of starting materials, the viscosity of a resin, and, indirectly, its solids content at a given viscosity, can be

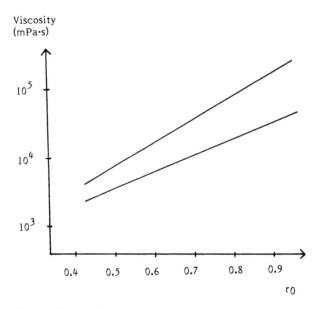

Figure 2.3 Viscosity as a function of the ratio of COOH groups to OH groups, r_0, for two linear polyesters.

predetermined by the choice of r_0. This type of reasoning will prove helpful in the design of high solids alkyd resins, although one must keep in mind that the resin synthesis is usually a complex process far away from an ideal polyesterification reaction.

For air-drying alkyds a simple reduction of the molecular weight, e.g., by adjusting r_0, in order to raise the solids content normally leads to inferior drying and film properties. Too much of the necessary cross-linking will have to be carried out by the autoxidation, a process which is more difficult to control than polymer synthesis in a reactor.

For stoving alkyds used in combination with amino resins the situation is somewhat different. A reduction of the alkyd molecular weight in order to achieve higher solids content of the paint seems to be more feasible in this case since the curing process can be relatively well controlled by the choice of amino resin, as well as by selection of catalyst. In general, the lower the molecular weight of the alkyd, the more critical is the choice of the amino resin component and also the ratio of alkyd to amino resin [7-10]. The amount of amino resin needed to obtain proper curing is normally considerably higher for high solids than for conventional systems.

Moreover, it has been found that the molecular weight of the alkyd influences the formation of the network and the type of structural elements formed in the film. In a study using hexa-methoxymethylmelamine as amino resin and three different polyesters, all based on cyclohexanedimethanol, phthalic acid, and adipic acid, the effect of a change in $\overline{DP}$ of the polyester was investigated [11]. The experiments were performed at a constant polyester-to-melamine resin ratio, and the stoving conditions were 35 min at 130°C.

As can be seen from Figure 2.4, the conversion of poly-ester, i.e., its chemical incorporation into the network, increased with increasing $\overline{DP}$. The reverse was found for the conversion of the amino resin. This behavior is the expected one and it reflects the decrease in ratio between reactive groups on the polyester and on the melamine resin (mainly OH and CH_2OCH_3 groups, respec-tively) when increasing the $\overline{DP}$ of the polyester. Also the degree of self-condensation of the melamine resin increased significantly with increasing polyester molecular weight. IR measurements re-vealed that in the film formed from the polyester having the low-est $\overline{DP}$, only 15% of the reactive groups of the melamine resin had been used for self-condensations. As the $\overline{DP}$ of the polyester increases, the amount of cocondensation between the two resin components decreases and the domains of melamine resin homo-condensates increases. In the curing of the polyester having the

Conversion

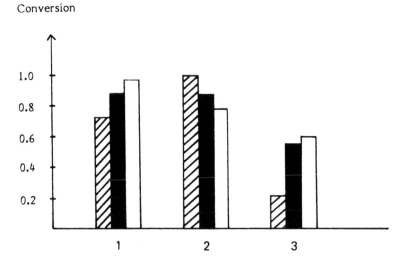

Figure 2.4 Conversion of a polyester (1) and hexamethoxy-methylmelamine (2), as well as self-curing of a hexamethoxy-methylmelamine (3). Three different polyesters were used: □ : DP 23, ■ : DP 11 and ▨: DP 3. Curing conditions are 35 min at 130°C. (From Ref. 11.)

highest DP about 65% of the reactive groups of the melamine resin had taken part in self-condensation. Thus, in the curing with melamine resins a low molecular weight alkyd resin gives a more uniform cross-linking and distribution of the components than a higher molecular weight alkyd. In practice, however, the difference in resin distribution is not so pronounced since the higher functionality of the low molecular weight polyester is normally compensated for by a somewhat higher melamine-to-polyester ratio.

Alkyd-melamine resin curing is dealt with in detail in Chapter 5.

2.2 MOLECULAR WEIGHT DISTRIBUTION

The molecular weight distribution was early recognized as one of the major parameters controlling the properties of polymers, both in solid form and in solution. In the coatings area there has always been a desire for better control of the polymerization process in order to arrive at polymers with a better defined structure and a more narrow molecular weight distribution. Group transfer polymerization seems to represent a major breakthrough in this respect in the field of acrylate and methacrylate polymerization [12,13]. In the synthesis of alkyds/polyesters no tool is to date available to provide such a high degree of control over polymer architecture. The step-growth method for alkyd preparation (see Section 3.4) is an example of an endeavor in the same direction, however.

2.2.1 Chromatographic Separation of Molecular Weight Fractions

The weight average molecular weight, $\overline{M_w}$, of a polymer is always higher than the number average, $\overline{M_n}$. The ratio $\overline{M_w}/\overline{M_n}$ is normally used as a measure of the distribution of the molecular weights of the individual polymers. Alkyds normally have a very broad molecular weight distribution, with the value of $\overline{M_w}$ 10-100 times that of $\overline{M_n}$.

The measurement of molecular weight distribution used to be a time-consuming and tedious procedure. Separation techniques were employed based on fractional extraction with solvents of different polarity or on fractional precipitation with mixtures of solvents and nonsolvents [14,15]. With the introduction of gel permeation chromatography (GPC), a technique that became established within the coatings industry during the 1970s, a chromatogram showing the molecular weight distribution can be obtained in a short period of time. GPC is used not only as an analytical tool but also to give preparative fractionation of the polymer. The chromatogram normally gives amount of material (detection by refractive index or UV measurement) as a function of elution volume. The largest molecules spend less time in the porous structure of the column and are eluted first. The elution volumes are transformed into values of molecular weight by use of a polystyrene standard.

Figure 2.5 shows gel permeation chromatograms of one conventional and one high solids model alkyd [16]. The latter resin exhibits an unusually narrow molecular weight distribution.

By using preparative GPC an alkyd can be sliced into an arbitrary number of molecular weight fractions, and each fraction

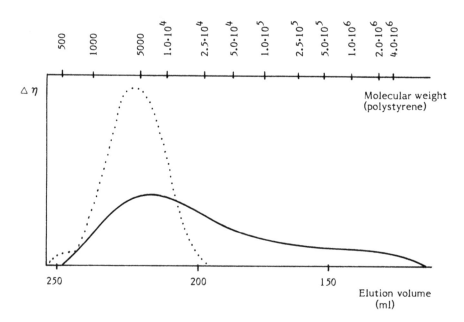

Figure 2.5 Molecular weight distribution of one conventional alkyd (full line) and one high solids alkyd (dotted line). Detection was made by refractive index, η, measurements given as $\Delta\eta$ on the ordinate. (From Ref. 16.)

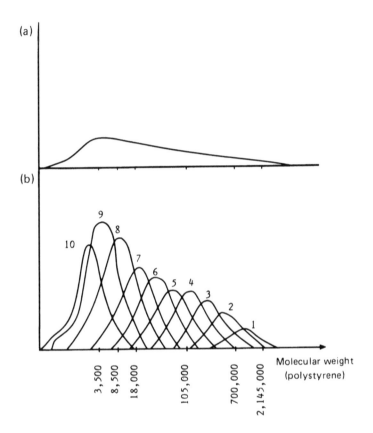

Figure 2.6   (a) Molecular weight distribution of a medium oil length, air-drying alkyd resin. (b) Gel permeation chromatograms of individual fractions of the original alkyd. (From Ref. 17.)

can be regarded as a resin of its own with well-defined $\overline{M_w}$ and $\overline{M_n}$ values and with characteristic properties. A commercial, medium oil length, air-drying alkyd resin was fractionated using preparative GPC [17]. Figure 2.6 shows the chromatogram of the original alkyd and the individual chromatograms of the 10 fractions collected. Table 2.1 gives numerical values for $\overline{M_w}$, $\overline{M_n}$, and $\overline{M_w}/\overline{M_n}$ of the fractions. It is apparent from the table that an effective

Table 2.1 $\overline{M_w}$ and $\overline{M_n}$ Values of a Medium Oil Length, Air-Drying Alkyd and of Its Individual Molecular Weight Fractions

Resin or fraction	Weight percent	$\overline{M_w}$	$\overline{M_n}$	$\dfrac{\overline{M_w}}{\overline{M_n}}$
Original resin	100	254,000	4,700	54
F1	2.7	1,180,000	1,000,000	1.2
F2	4.8	580,000	390,000	1.5
F3	5.4	294,000	203,000	1.4
F4	6.5	226,000	105,000	2.2
F5	7.5	148,000	49,000	3.0
F6	8.9	39,000	27,000	1.5
F7	12.3	21,000	13,000	1.6
F8	16.9	10,000	5,800	1.7
F9	17.9	4,900	2,650	1.9
F10	11.6	2,720	1,660	1.6

fractionation can be obtained over the very wide molecular weight span.

GPC has proved to be a useful tool to monitor small changes in alkyd structure as a result of variations in the ratio of starting materials or in the synthesis procedure [18-20]. Later versions of the technique are sometimes referred to as HPGPC (High Performance GPC) [18].

2.2.2 Role of Individual Molecular Weight Fractions

Analyses of each fraction obtained from preparative GPC of alkyd resins show surprisingly large differences both in acid value and hydroxyl number between the fractions. In two separate investigations it was found that the acid value passed through a minimum at a molecular weight of a few thousand [17,20]. Also, it was shown that the chemical composition of the fractions differed considerably, particularly in the lower molecular weight range [17]. Evidently, ester bond cleavage, ring closure, and other side reactions cause deviations from the normal Flory theory for poly-esterification (see Chapter 3).

Figure 2.7 shows the solids content at a given viscosity for the various fractions obtained by preparative GPC of a medium oil length, air-drying alkyd resin [17]. As expected, solvent de-mand increases with increased molecular weight.

The higher molecular weight fractions are evidently un-wanted from a solids content point of view. For air-drying sys-tems a certain proportion of higher molecular weight material is needed, however, in order to obtain proper drying of the film. Figure 2.8 shows the drying times for the individual fractions from the above-mentioned alkyd. As can be seen, the lower molecular weight fractions exhibit extremely poor drying properties. Frac-

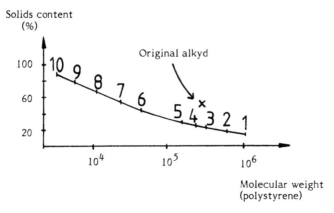

Figure 2.7 Solids content of alkyd fractions in xylene at a con-stant viscosity of 370 mPa · s at 23°C. (From Ref. 17.)

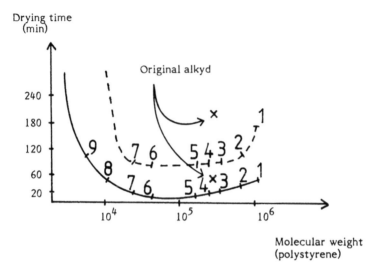

Figure 2.8 Drying time of alkyd fractions dissolved in mineral spirits/xylene 3:1 and with Co, Pb, Zr drier added. Full line: dry-to-touch; dashed line: through-dry. (From Ref. 17.)

tions F9 and F10 are not through-dry after one week. The increase in drying time at the high molecular end can probably be attributed to solvent retention; during the drying of the very high molecular weight fractions, solvent is entrapped in the coating (see Section 4.2).

Figure 2.8 is a good illustration of the statement made earlier that a simple reduction of the molecular weight of the alkyd resin is not a practicable way to obtain high solids coatings, since the increase in solids content will be accompanied by a severe reduction of drying capability.

In general, it is known that the low molecular components have a positive influence on adhesion, flexibility, and gloss. The high molecular components are responsible for quick drying, high hardness, and good solvent resistance [16]. It has also been claimed that the high molecular part can be considered as the film-forming binder, while the low molecular part serves as a combined solvent and plasticizer [21]. A compromise between a high solids content and good drying and film properties can be obtained by cutting off both tails of the molecular weight distribution curve. For the above mentioned air-drying alkyd this has been performed by recombining fractions 6, 7, and 8, obtained from preparative GPC (see Table 2.1 and Fig. 2.7). This new alkyd has an $\overline{M_w}/\overline{M_n}$ of 4.0 and is practically void of molecules below 2,000 and above 200,000 molecular weight. Its drying and film properties were at least equal to those of the original alkyd.

Cutting of the tails of the molecular weight distribution curve is hardly feasible in modern large-scale operations. (Although technically feasible examples of selective extractions, as well as of incremental precipitations, of molecular weight fractions appear in the literature [22,23].) The clue to high solids alkyd resins is to find a process for the preparation that gives sufficient control over the architecture of the polymer.

2.3 BRANCHING

The solution viscosity of a polymer depends on the degree of branching. Branched polymers normally give lower viscosities than their linear counterparts of equal molecular weight [24]. Large, "bushy" branches seem to be particularly effective in this respect [25]. The investigations of the effect of polymer structure on viscosity have been performed mainly with carefully designed addition polymers.

It is difficult to synthesize alkyd resins with a well-defined branching of the skeleton. It is also difficult to achieve such control over the polymerization process that parameters such as molecular weight distribution, acid value, and hydroxyl number, which all affect the viscosity, are kept constant while varying the degree of branching. Therefore, attempts to give a quantitative estimation of the effect of branching on viscosity of alkyd solutions seem to be lacking in the literature.

2.4 FUNCTIONAL GROUPS

2.4.1 Polar Groups

Polar groups in alkyds influence the solution viscosity to a large extent. Hydroxyl and carboxyl groups function as hydrogen bond donors, while ester and carboxyl groups are acceptors. A high con concentration of these functional groups in combination with a steric accessibility results in strong intermolecular forces and, thus, in a high viscosity. The strong viscosity-reducing effect exerted by small amounts of low-molecular ketones and alcohols, although their solubility parameters lie far from those of the binder, is partly due to the fact that they favorably compete with the alkyd as acceptor and donor, respectively, of hydrogen bonds. (See also Section 4.3.)

In most cases, the concentration of carboxyl groups is relatively low and the intermolecular associations are therefore mainly caused by hydroxyl and ester groups. There is not much to be done about the latter groups—by definition, the alkyd is a polyester. The hydroxyl group concentration is also a relatively fixed parameter. Baking alkyds require a certain hydroxyl number to react with the amino resins, and for air-drying alkyds the concentration of hydroxyl groups determines good drying and pigment wetting properties.

In order to study the effect of residual hydroxyl groups on the viscosity of alkyd solutions, a series of model alkyds with

varying hydroxyl number have been synthesized [19]. The resins
were built up by step-wise polymerization in order to get as good
control over the polymer structure as possible. Care was taken
that the acid value and the hydroxyl number, as well as the basic
chemical composition, were roughly the same for all members of
the series. Polyols of different functionality were used in order
to achieve the various hydroxyl numbers of the alkyds. However,
the polyols used were of similar type, containing only primary
alcohols with the hydroxymethyl groups attached to a tertiary
carbon atom. Hence, the hydrogen bonding ability of each hydroxyl group can be expected to be approximately the same.

In Figure 2.9 the viscosities of the alkyds in 80% solutions in xylene are given as a function of the number of OH/mol
[19]. In the same diagram results from another study are shown,
where the residual hydroxyl groups of an alkyd have been partly
or fully capped by reaction with acetic anhydride [16]. The influence of hydroxyl groups on solution viscosity is here expressed
as viscosity vs. hydroxyl number.

As expected, the viscosity of the alkyd resin increased
with increasing concentration of hydroxyl groups. It can be seen,
however, that the slopes of the curves are not very steep and that
a considerable reduction in hydroxyl number is needed in order
to get a substantial viscosity decrease. A full, reversible, blocking
of the polar groups, as discussed in Section 3.5, is one way to
achieve this.

2.4.2 Olefinic Groups

Unsaturated groups in the fatty acid residues of air-drying alkyds
exert a relatively strong effect on the viscosity of alkyd solutions.
This was demonstrated in a recent work where the solution viscosities of a series of alkyd resins, similar in structure, acid value,

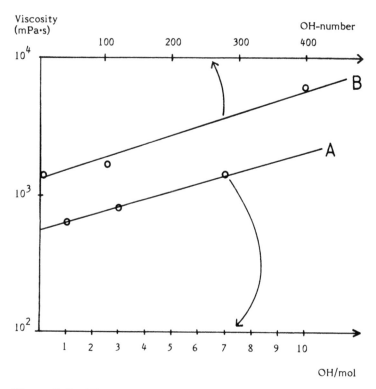

Figure 2.9 Viscosity of two series of alkyd resins as a function of (A) OH/mol and (B) OH number.

and hydroxyl number and differing only in the concentration of olefinic groups, were determined [19]. The variation in unsaturation was attained by the use of fatty acids having varying number of olefinic groups. The result is shown in Figure 2.10. The effect of fatty acid unsaturation on viscosity is probably related to a reduction of the intermolecular associations. The increased rigidity of the fatty acid chains caused by the olefinic groups is likely to render the formation of intermolecular forces more difficult.

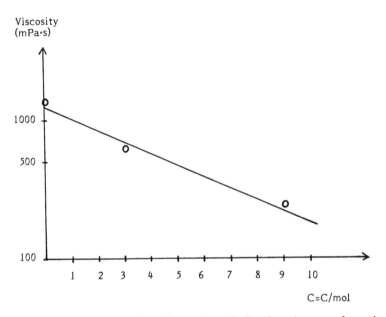

Figure 2.10 Viscosity of a series of alkyd resins as a function of the number of C = C/mol.

REFERENCES

1. L. W. Hill, *Prog. Org. Coat.*, *10*:55 (1982).

2. A. Mercurio and S. N. Lewis, *J. Paint Technol.*, *47*, No. 607: 37 (1975).

3. C. K. Schoff, *Prog. Org. Coat.*, *4*:189 (1976).

4. D. Stoye and J. Dörffel, *Proc. 4:th Intern. Conf. Org. Coat. Sci. Technol.*, Athens, Greece, 1978, p. 349.

5. D. Stoye and J. Dörffel, *Pigm. Resin Technol.*, *9*:4 (July 1980).

6. D. Stoye and J. Dörffel, *Pigm. Resin Technol.*, *9*:8 (August 1980).

7. J. Dörffel, *Farbe Lack*, *88*:6 (1982).

8. W. J. Blank, *J. Coat. Technol.*, *54*:26 (1982).

9. N. J. Albrecht and W. J. Blank, *Adv. Org. Coat. Sci. Technol. Ser.*, *4*:1 (1982).

10. S. N. Belote and W. W. Blount, *J. Coat. Technol.*, *53*:33 (1981).

11. D. Stoye and J. Dörffel, *Org. Coat. Sci. Technol.*, *6*:257 ((1984).

12. O. W. Webster, W. R. Hertler, D. Y. Sogah, W. B. Farnham, and T. V. Rajan Babu, *J. Amer. Chem. Soc.*, *105*:5706 (1983).

13. C. A. Senkler, *Org. Coat. Sci. Technol.*, *8*:1 (1986).

14. J. Ivánfi, *Farbe Lack*, *70*:426 (1964).

15. H. Neises, *Farbe Lack*, *77*:850 (1971).

16. H.-J. König, *Farbe Lack*, *84*:756 (1978).

17. G. Walz, *J. Oil Col. Chem. Assoc.*, *60*:11 (1977).

18. C.-Y. Kuo, T. Provder, and A. F. Kah, *Org. Coat. Sci. Technol.*, *6*:101 (1984).

19. K. Holmberg and J.-A. Johansson, *Org. Coat. Sci. Technol.*, *6*:23 (1984).

20. J. Kumanotani, H. Hata, and H. Masuda, *Org. Coat. Sci. Technol.*, *6*:35 (1984).

21. T. R. Bullett and A. T. S. Rudram, *J. Oil Col. Chem. Assoc.*, *42*:778 (1959).

22. H. J. Wright and R. N. Du Puis, *Ind. Eng. Chem.*, *36*:1004 (1944).

23. D. H. Solomon, *The Chemistry of Organic Film Formers*, Krieger, New York, 1977, p. 105.

24. C. E. H. Bawn and M. B. Huglin, *Polymer*, *3*:615 (1962).

25. J. A. Manson and L. H. Cragg, *Can. J. Chem.*, *36*:858 (1958).

3

Principles for the Preparation of High Solids Alkyds

3.1 THEORY OF POLYESTERIFICATION

3.1.1 General

Berzelius, reacting glycerol with tartaric acid, was probably the first to produce a synthetic, cross-linked polyester [1]. The first linear polyester with a defined structure was prepared some 80 years later by Carothers [2]. Flory and Carothers are generally recognized to have laid the basis for the theory of polyesterification.

Polyesters may be prepared either by direct condensation of at least difunctional acids and at least difunctional alcohols, or by employing reactive derivatives of the acid or the alcohol component. The first type of reaction is reversible, and in order to shift the equilibrium towards the product side, the water formed must be removed from the reaction zone. In practice, such removal may be carried out in various ways, such as azeotropic distillation using an organic solvent, sweeping the vapor away by means of a stream of inert gas, or by applying a vacuum.

Reactions with reactive derivatives may be regarded as irreversible. Anhydrides and epoxides are the most common types of reactive derivatives of acids and alcohols, respectively. The possibility of using other types of derivatives is discussed in Section 3.4.

Polyesterification is one of the prime examples of step-growth polymerization. This is a type of reaction in which each polymer chain grows at a relatively slow rate over a much longer period of time than in an addition polymerization reaction, and in which the initiation, propagation, and termination reactions are approximately identical in both rate and mechanism [3].

The equilibrium constant of polyesterification is normally equal to that of the analogous model reaction between monofunctional compounds. This has been explained by the proposition that the reactivity of the functional end groups in the growing polyester chain is independent of the degree of polymerization. In other words, at all stages of polymerization the reactivity of every functional group is the same. This principle of equal reactivity of functional groups, first demonstrated by Flory [4,5], has been of tremendous importance in the synthesis of alkyds since it permits the application of statistical considerations to the problem of the distribution of the bonds formed during polymerization.

The principle of equal reactivity also means that each monomer unit can react with other monomer units or with polymer chains with equal ease. This is why the polymer chain grows at a comparatively slow rate over a long period of time.

The random growth of a polymer by the step-growth mechanism permits only a very gradual increase in molecular size. In the polyesterification of a hydroxy acid, e.g., 4-hydroxymethyl-benzoic acid, as shown below, or of equimolar amounts of a glycol and a dibasic acid, a conversion of 75% of the functional groups gives an average degree of polymerization of only five. A conversion of 95% corresponds to a degree of polymerization of 20, in both cases assuming ideal conditions.

$$n \cdot HOCH_2 - \!\!\bigcirc\!\! - COOH \longrightarrow \left[OCH_2 - \!\!\bigcirc\!\! - \overset{\overset{\displaystyle O}{\|}}{C} \right]_n + n \cdot H_2O$$

3.1.2 Degree of Polymerization: Carothers Equation

In the following discussion $\overline{DP}$ is an abreviation for the number average degree of polymerization, i.e.,

$$\overline{DP} = \frac{\text{monomers in the system}}{\text{molecules in the system}}$$

In the polycondensation, A groups, e.g., carboxyl groups, react with B groups, e.g., hydroxyl groups. The reactants are assumed to be bifunctional, and the two functional groups of each molecule may be the same or different. Hence, not only dicarboxylic acids (A—A) and glycols (B—B) are accounted for. (Note that A and B mean groups, not molecules.)

Initially, the number of moles of A is $(n_A)_0$ and of B $(n_B)_0$. The number of A groups is always smaller than the number of B groups, i.e.,

$$r_0 = \left[\frac{n_A}{n_B} \right]_0 < 1$$

The extent of reaction, P_A, is defined as the fraction of A groups (since A is the limiting functional group) that have reacted at a given conversion. No side reactions are accounted for.

After a certain conversion the number of moles of A groups is

$$n_A = (n_A)_0 - P_A \cdot (n_A)_0 \tag{3.1}$$

Assuming only reactions between A and B groups:

$$n_B = (n_B)_0 - P_A \cdot (n_A)_0 \tag{3.2}$$

At a specific conversion, the total amount of end groups is

$$n_A + n_B = (n_A)_0 - P_A \cdot (n_A)_0 + (n_B)_0 - P_A \cdot (n_A)_0 \tag{3.3}$$

which after transformation gives

$$n_A + n_B = (n_A)_0 \left[2(1 - P_A) + \frac{1 - r_0}{r_0} \right] \tag{3.4}$$

By definition:

$$\overline{DP} = \frac{n_{monomers}}{n_{molecules}} \tag{3.5}$$

For bifunctional reactants the number of molecules is half the number of end groups. Hence,

$$n_{monomers} = \frac{(n_A)_0 + (n_B)_0}{2} \tag{3.6}$$

and

$$n_{molecules} = \frac{n_A + n_B}{2} \tag{3.7}$$

Combining Eq. (3.6) and (3.7) with Eq. (3.5) gives

$$\overline{DP} = \frac{(n_A)_0 + (n_B)_0}{n_A + n_B} \tag{3.8}$$

or, with Eq. (3.4) and after transforming:

$$\overline{DP} = \frac{r_0 + 1}{2r_0(1 - P_A) + 1 - r_0} \tag{3.9}$$

In alkyd synthesis the alcohol component usually is used in excess. Hence, carboxyl groups (A) are present in lower concentrations than hydroxyl groups (B) and, consequently, $r_0 < 1$. When the reaction goes towards completion, P_A approaches 1 and Eq. (3.9) becomes

$$\overline{DP} = \frac{r_0 + 1}{1 - r_0} \tag{3.10}$$

This equation is of considerable practical importance. It shows that the maximum degree of polymerization that can be obtained is given by the initial molar ratio of functional groups, i.e., for polyesterification the ratio of carboxyl to hydroxyl groups. The closer the ratio is to unity, the higher will be the molecular weight. In practice, r_0 above 0.90 is always employed. The dependence of $\overline{DP}$ on r_0, calculated from Eq. (3.10), is given in Table 3.1.

It is clear from Eq. (3.9) that the more equimolar the reaction mixture (r_0 approaching 1), the higher the degree of polymerization. When equimolar initial concentrations are used, Eq. (3.9) reduces to

$$\overline{DP} = \frac{1}{1 - P_A} \tag{3.11}$$

The degree of polymerization is now only dependent on the extent of reaction. At a reaction extent of 100%, $\overline{DP}$ is infinite.

Table 3.1   Dependence of the
Maximum Degree of Polymerization, DP_{max}, on the Initial Molar
Ratio of Functional Groups, r_0,
at 100% Extent of Reaction

r_0	$\overline{DP}_{max}$
0.5000	3
0.9000	20
0.9900	200
0.9990	2,000
0.9999	20,000

In commercial alkyd synthesis ideal reactions between bifunctional reagents never occur. First of all, starting materials of
both higher and lower functionality than two are often employed.
Furthermore, various side reactions, as well as the fact that the
actual reactivity of the functional groups depends on their chemical environment, must be taken into account. Nevertheless, the
theory of polyesterification leading to Eq. (3.11), which is sometimes referred to as Carothers' Equation, plays an important role
in modern alkyd synthesis. In Section 3.1.4 the principle of calculations of alkyd synthesis, including treatment of starting materials with functionalities other than two, is given. These calculations are normally based on Carothers' theory. Side reactions
in alkyd synthesis are dealt with in depth in Section 3.2.

3.1.3   Molecular Weight Distribution

The molecular weight distribution has in recent years been recognized as being one of the most critical parameters of alkyd resins.
It has become increasingly clear that the physical properties of an
alkyd can be said to represent the sum of the contributions of

the various molecular weight fractions. With special regard to high solids alkyds it is clear that a narrow molecular weight distribution is a necessary condition since the higher molecular weight fraction requires too much solvent in order to obtain acceptable viscosity and the lower weight fraction gives poor film properties. This topic is discussed in more detail in Section 2.2.

In order to calculate the amount of unreacted monomer at any stage in a step-growth polymerization, one must use an equation derived from probability considerations. The approach used is to determine the probability of selecting at random a molecule with exactly x repeating units. This probability is equivalent to the mole fraction, N_x, of polymer chains with x repeating units. The probability that one type of functional group has reacted is called p and, consequently, the probability that it has not reacted is $1 - p$. It can now be shown [6] that, for equimolar initial concentrations of reactants,

$$N_x = N_0(1 - p)^2 \cdot p^{x-1} \tag{3.12}$$

Likewise, for the weight fraction, w_x, of polymer chains with x repeating units it can be shown that

$$w_x = m(1 - p)^2 \cdot p^{x-1} \tag{3.13}$$

where m is the molecular weight of the repeating unit.

The mole fraction and weight fraction distributions of polymer chain lengths, x, according to Eqs. (3.12) and (3.13), respectively, are plotted in Figures 3.1 and 3.2. The diagrams give some indication of the extremely broad distribution of molecular species formed in a step-growth polymerization reaction. The higher the degree of polymerization, DP, the smaller will be the mole fraction, N_x. The corresponding weight-fraction, w_x, distribution curves, on the other hand, pass through maxima.

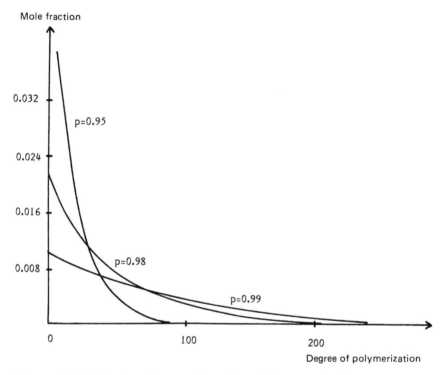

Figure 3.1 Mole fraction as a function of degree of polymerization for a polycondensation reaction. p = extent of reaction.

As the average degree of polymerization increases, the weight fraction distribution becomes extremely broad.

Figures 3.1 and 3.2 are useful in visualizing the product
mixture of a step-growth reaction, such as a polyesterification.
They show that the molecular weight distribution becomes
broader as the extent of reaction approaches 1. If, for instance,
a certain polymer preparation needs a DP of 50 in order to reach
acceptable product properties, an extent of reaction of at least
0.98 is needed. Actually, very few condensation reactions proceed
to such a high yield even when difunctional reactants are being

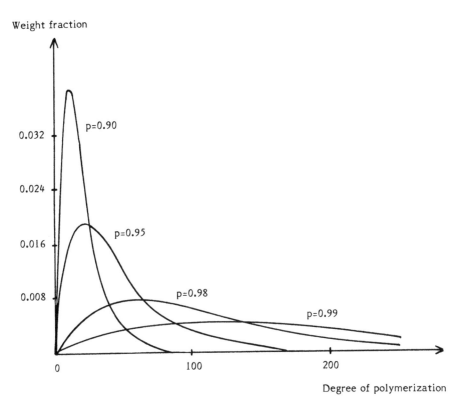

Figure 3.2   Weight fraction as a function of degree of polymerization for a polycondensation reaction. p = extent of reaction.

used. Polyesterifications are among only three of four types of step-growth condensations of industrial importance.

As was discussed in Chapter 2, one of the keys to making high solids alkyds is to synthesize polymers with a narrow molecular weight distribution. As seen in Figure 3.2, a relatively low extent of reaction is then necessary if the reaction proceeds according to a step-growth mechanism. The $\overline{DP}$ and, hence, the molecular weight will then be low and, as can be seen from Figure 3.1, the monomers and the lowest oligomers will be the molecules most abundant in the reaction mixture. This large fraction of low molecular material will be expected to cause problems in the curing process and to influence the final film properties in a negative way. This is one of the major dilemmas in the preparation of high solids alkyd resins.

3.1.4 Multifunctional Reactants: Actual Functionality

If, in a polyesterification, equimolar amounts of a glycol and a dibasic acid are reacted, both reactants are able to achieve their maximum functionality of two and an infinite polymer having both hydroxyl and carboxyl end groups will be obtained:

$$n \cdot \text{HO}-\text{R}-\text{OH} + n \cdot \text{HOOC}-\text{R}^1-\text{COOH}$$

$$\downarrow -(2n-1)\text{H}_2\text{O}$$

$$\text{HO}-\text{R}-\text{O}-\left[\text{CO}-\text{R}^1-\text{COO}-\text{RO}-\right]_{n-1}\text{CO}-\text{R}^1-\text{COOH}$$

If, on the other hand, two moles of a glycol is reacted with one mole of a dibasic acid, then, on the average, a hydroxyl terminated diester will be formed:

$$2\,\text{HO}-\text{R}-\text{OH} + \text{HOOC}-\text{R}^1-\text{COOH}$$

$$\downarrow -2\text{H}_2\text{O}$$

$$\text{HO}-\text{R}-\text{OCO}-\text{R}^1-\text{COO}-\text{R}-\text{OH}$$

In this case the actual functionality of the glycol is one, i.e., half its maximum or potential functionality. The acid is employed at its maximum functionality, i.e., its actual functionality is two. In actual practice, at least a slight deviation from equimolar amounts of reactive functional groups are always used.

In the synthesis of alkyds, alcohols and acids with functionalities of from one to four are commonly used as raw materials. The monofunctial reactants cause chain termination, whereas those with functionalities of more than two give rise to polymers that can have branched and eventually three-dimensional structures. The polymerization of these more highly functional systems follows the same general principles as the simpler bifunctional ones. The reactions proceed in a stepwise manner and the molecular weight distributions become broader with increasing degree of polymerization.

In the condensation of bifunctional reactants, the functionality of the reaction product is always two, regardless of the extent of the reaction. In the reaction between a triol and a dibasic acid, using equivalent amounts of hydroxyl and carboxyl groups, on the other hand, the functionality of the molecules formed increases as the reaction proceeds. This is illustrated below for the reaction between glycerol and adipic acid; the tetraester produced has a functionality of four.

$$2CH_2-CH-CH_2 + 3HOOC-(CH_2)_4-COOH \longrightarrow$$
$$\quad\; | \qquad | \qquad |$$
$$\quad OH \quad OH \quad OH$$

$$CH_2-CH-CH_2-OCO-(CH_2)_4-COO$$
$$| \qquad | \qquad\qquad\qquad\qquad\qquad |$$
$$OH \quad OH \qquad\qquad\qquad\qquad\quad CH_2$$
$$\qquad\qquad\qquad\qquad\qquad\qquad\qquad |$$
$$\qquad HOOC-(CH_2)_4-COO-CH$$
$$\qquad\qquad\qquad\qquad\qquad\qquad\qquad |$$
$$\qquad\qquad\qquad\qquad CH_2-OCO-(CH_2)_4-COOH$$

In this case the actual functionality equals the maximum functionality of both the triol and the dibasic acid; a highly cross-

linked structure will form and gelation will eventually occur.
(In reality, due to the relatively low reactivity of the secondary
hydroxyl group, the actual functionality of glycerol should always be calculated as being somewhat lower than its theoretical
value. This will be further discussed later.)

    If the glycerol and the adipic acid is, instead, used in
equimolecular amounts, a linear polymer will be obtained. The
actual functionality of the triol is now two, instead of three,
since on the average only two hydroxyl groups from each glycerol
molecule will react.

$$CH_2-CH-CH_2-O-[CO-(CH_2)_4-COO-CH_2\,CH-CH_2-O]-CO-(CH_2)_4-COOH$$
$$\underset{OH\quad OH}{} \qquad\qquad\qquad\qquad \underset{OH\qquad\quad n}{}$$

    In most alkyd syntheses an excess of hydroxyl over car-
boxyl groups is used. The actual functionality of the acid component is then equal to its maximum functionality. The actual
functionality of the polyol, on the other hand, is lower than the
maximum functionality and can be calculated from the following
formula [7]:

$$F_{actual} = \frac{F_{maximum}}{1+n}$$

where n is the fraction of hydroxyl groups present in excess of
carboxyl groups.

    For example, in the above case where glycerol and adipic
acid are used in equimolecular amounts hydroxyl groups are
present in 50% excess. Hence, the actual functionality will be:

$$F_{glycerol} = \frac{3}{1+0.5} = 2$$

Often mixtures of polyols with varying functionality are employed. A mean (maximum) functionality of the polyol mixture is then calculated using the formula:

$$F_{polyol} = \frac{\text{total equivalents OH}}{\text{total moles polyol}} = \frac{N_A \cdot F_A + N_B \cdot F_B + \cdots}{N_A + N_B + \cdots}$$

where

N_x = the number of moles of polyol X

F_x = the maximum functionality of polyol X

The same procedure is used to calculate the mean functionality of a mixture of acids. Esters, which in the form of triglycerides are common raw materials, are assigned zero functionality. Mono- and diglycerides are, of course, regarded as alcohols having functionalities of two and one, respectively. A hydroxy acid, such as 4-hydroxymethylbenzoic acid (see p. 49), should be calculated as both an acid of functionality one and an alcohol of the same functionality.

Now, the overall maximum functionality, including both acid and alcohol components, is expressed as

$$F_{overall,max} = \frac{\text{total equivalents}}{\text{total moles}}$$

A more useful expression is obtained if we disregard the excess of one component (usually excess polyol):

$$F_{overall,act} = \frac{\text{total equivalents} - \text{excess equivalents}}{\text{total moles}}$$

The following example illustrates the procedure. An alkyd was prepared according to the formulation of Table. 3.2.

Total equivalents polyols: $1.0 + 4.5 = 5.5$
Total equivalents acids: $0.8 + 3.6 = 4.4$
Excess equivalents polyol: $5.5 - 4.4 = 1.1$

Table 3.2 Alkyd Formulation

Raw materials	Moles	Max functionality	Equivalents
Diethylene glycol	0.5	2	1.0
Glycerol	1.5	3	4.5
Fatty acid	0.8	1	0.8
Adipic acid	1.8	2	3.6

$$F_{overall,act} = \frac{(5.5 + 4.4) - 1.1}{4.6} = 1.91$$

Now, the overall actual functionality can be used as a measure of whether an alkyd composition will eventually gel or not. In doing this, Carothers' definition of gel point, i.e., that gelation corresponds to an infinite number average molecular weight, is normally used [8-11].

The extent of reaction, P, is then defined as the fraction of functional groups reacted, i.e.,

$$P = \frac{2(N_0 - N)}{N_0 \cdot F_{overall,act}}$$

where

N_0 = the number of molecules initially present

N = the number of molecules present after reaction

The expression $2(N_0 - N)$ then represents the number of functional groups that has reacted.

Since at the gel point $N_0 \gg N$, the extent of reaction at gel can be written:

$$P_{gel} = \frac{2}{F_{overall,act}} \qquad\qquad (3.14)$$

Now, if P_{gel} is < 1, the composition will gel before reaction has gone to completion. A value of $P_{gel} > 1$ indicates that the reaction will not gel (assuming ideal conditions).

Using the value of $F_{overall,act}$ obtained from the values of Table 3.2 above gives

$$P_{gel} = \frac{2}{1.91} = 1.05$$

This indicates that 105% conversion is needed to give gelation. Consequently, the composition will not gel.

The above theory has been the basis for computerized calculations on alkyd formulations. It is well known that alkyds should be processed close to the gel point; gelation during cooking is a disaster, however, even if a method of saving such runs has recently been proposed [12]. In the calculation the relative amounts of the ingredients can be varied so as to arrive at a gel point at a certain acid value. Furthermore, the formulation can be modified by incorporating other acids and alcohols into the recipe, as well as excluding original ones. The only demand that is being made on a new raw material is that its molecular weight and maximum functionality are known.

The calculations can easily give not only the relative amounts of the starting materials to arrive at gel point at a certain acid value, but also the molecular weight of the alkyd at a given acid value, higher than the acid value corresponding to gelation. This has been found to be extremely useful, not least in the design of high solids alkyds, where the molecular weight of the polymer is a crucial parameter.

The gel point in the discussion above corresponds to an infinite number average molecular weight. Alternatively, if the gel point is assumed to be correlated to an infinitely large weight average molecular weight, it can be shown that [13,14], assuming an excess of hydroxyl groups,

$$P_{gel} = \sqrt{\frac{\epsilon}{2(1 - \lambda)}}$$

(3.15)

where

$$\epsilon = \frac{\text{equivalents of hydroxyl}}{\text{equivalents of carboxyl}}$$

$$\lambda = \frac{\text{equivalents of monobasic acid}}{\text{total equivalents of acid}}$$

Either Eq. (3.14) or (3.15) can be used to optimize an alkyd formulation. In practice, the theory based on number average molecular weight is most often used for calculating alkyd syntheses. One reason for this is that the normal way of monitoring the molecular weight growth during synthesis is by end-group analysis, i.e., determination of the concentrations of hydroxyl and carboxyl groups. Since this is a measure of the number average molecular weight of the polymer, the method of analysis is directly correlated to Eq. (3.14). It should be kept in mind that the number average molecular weight is always considerably lower than the weight average (unless all molecules are of the same size) and that the influence on viscosity, which is of paramount interest for high solids alkyds, is essentially related to the latter parameter.

The above-mentioned discrepancy between number average and weight average molecular weight is one reason the theoretical calculations on gel point must be used with care. Another reason is that the polyesterification is not an ideal process, but is subject to a number of side reactions which influence the structure

of the final polymer. Furthermore, the true functionality of the
starting materials may not always correspond to the theoretical
values since the reactivity of a functional group is very much dependent on its chemical environment.

The most important correction factors that have to be
taken into account in alkyd synthesis calculations are

1. The true functionality of a starting material is lower
than the theoretical value. The reason for this behavior is normally that the reactivity of one of the functional groups of a multifunctional reactant is reduced due to steric and/or electronic
effects. One example of this is glycerol, the secondary hydroxyl
group of which reacts much more sluggishly with carboxyl groups
than the two primary groups.

2. The true functionality is higher than the theoretical
value. This applies to polyols in general since etherifications invariably occur parallel to the esterification reactions (see Section
3.2). Etherification of a polyol having n hydroxyl groups leads to
a new polyol with a functionality of $2n - 2$, as illustrated below
for pentaerythritol:

$$2\ HOCH_2-\overset{\displaystyle CH_2OH}{\underset{\displaystyle CH_2OH}{C}}-CH_2OH \xrightarrow{-H_2O} HO-CH_2\cdots\overset{\displaystyle CH_2OH}{\underset{\displaystyle CH_2OH}{C}}-CH_2\cdots O-CH_2\cdots\overset{\displaystyle CH_2OH}{\underset{\displaystyle CH_2OH}{C}}-CH_2OH$$

In reality, commercial polyols often contain significant amounts
of the dimer. This is not necessarily a disadvantage as long as the
exact amount, and thus, the true functionality of the product, is
known.

Another common example of the true functionality being
higher than the theoretical value is that of unsaturated fatty acids.

Cross-linking between fatty acid chains is known to occur during alkyd processing, probably mainly as a result of thermal isomerization to structures containing conjugated double bonds followed by a Diels-Alder reaction between two chains [15,16]. This, of course, does not increase the functionality of the acid in the sense of it consuming more hydroxyl groups, but it does contribute to the degree of branching of the polymer network and, hence, it increases the risk of gelation.

   3. The theory predicts only intermolecular reactions. Intramolecular reactions will invariably reduce the true functionality. Intramolecular etherification of a polyol, which is quite feasible in the case of, for instance, diethylene glycol, reduces the functionality by 2. Intramolecular esterifications, leading to cyclizations, as is shown below for the monoester of glycerol and phthalic acid, also reduce the functionality.

These and other deviations from ideal conditions of polyesterification have led to the application of a number of correction factors to the various starting materials in alkyd synthesis. The situation is very complex, however, since the ratios between the rate constants of the various reactions involved, such as esterification, transesterification, etherification, Diels-Alder reaction, ene reaction, etc., vary with temperature and is also influenced by the use of an esterification catalyst. Computer-based calculations can, of course, easily handle the parameters involved, but the setting of the correction factors is still to a large extent based on trial and error.

3.1.5 Kinetics

Flory [5,17] and, later, Hamann et al. [18], carried out model polyesterification reactions to show that third-order kinetics were essentially followed if the reactions were performed in the absence of an added catalyst. The carboxyl groups act as catalyst and the mechanism involved is the following:

$$2\,RCOOH \rightleftharpoons RCOO^- + RCOOH_2^+ \qquad (3.16)$$

$$RCOOH_2^+ + R^1OH \rightleftharpoons RCOOHR^{1+} + H_2O \qquad (3.17)$$

$$RCOOHR^{1+} + RCOO^- \rightleftharpoons RCOOR^1 + RCOOH \qquad (3.18)$$

Reaction (3.17) is believed to be the rate-determining step [19]. In reaction media of low dielectric constant, such as esters and polyesters, the ions are probably associated as ion pairs. The decrease in concentration of carboxyl groups can be expressed:

$$-\frac{d[COOH]}{dt} = k \cdot [COOH]^2 \cdot [OH] \qquad (3.19)$$

The molar concentration $[COOH]$ at time t is related to the original molar concentration $[COOH]_0$ through the degree of reaction, P. For equimolar initial concentrations of reactants Eq. (3.11) expresses the relationship between P and degree of polymerization, $\overline{DP}$. Combining Eq. (3.11) and (3.19) gives, after integration:

$$\frac{1}{(1-P)^2} = 1 + 2k[COOH]_0^2 \cdot t = (DP)^2 \qquad (3.20)$$

A linear relationship between $1/(1-P)^2$ and t is found for values of P between 0.8 and 0.98 [6]. This corresponds to $\overline{DP}$ values of

5 and 50, respectively, thus covering the range of polymerization of interest for high solids alkyds.

If the polyesterification is performed in the presence of an acid catalyst, the reaction becomes second order over a wide range of $\overline{DP}$ [18,19]. At high degrees of conversion, however, the reactions become sluggish. This has been ascribed to depletion of the catalyst; at low concentration of remaining COOH groups a catalyst, such as *p*-toluene sulphonic acid, may compete favorably in reacting with OH groups, thus acting as a chain-terminating additive [17,20].

3.2 SIDE REACTIONS

In the preparation of alkyds, the polycondensation is accompanied by a number of side reactions and by secondary transformations of the products formed. In specifying an alkyd, it is necessary to know the approximate extent of these side reactions and to account for them in the formulation.

The relative importance of various side reactions depends to a large extent on the specific raw materials used in the synthesis. For instance, it is well known that *o*-phthalic acid gives rise to considerably more cyclic, chain-terminating structures than isophthalic acid and that neopentyl glycol, when esterified into the alkyd backbone, is substituted less readily by other alcohols (alcoholysis) than most other glycols used in alkyd preparation.

To a certain extent these deviations from ideal conditions can be quantified and accounted for in computer-based calculations of alkyd formulations. In many cases, however, a quantitative estimation of the importance of the side reactions are lacking, and the adjustment of the alkyd formulation must be made empirically.

This type of consideration is particularly important in the design of high solids alkyd resins since the tolerances in the preparation with regard to gelation, etc. is smaller for these than for conventional alkyds.

3.2.1 Etherification

The dehydration of alcohols to form ethers is usually an acid-catalyzed reaction, and the species from which the leaving group departs is ROH_2^+. If sulphuric acid is used as a catalyst, a mono-alkylsulphate ester may be an intermediate. The reaction scheme is shown in Figure 3.3.

Elimination, i.e., olefin formation, is always a possible side reaction. With primary alcohols, which are predominantly used in alkyd synthesis, elimination is probably of minor importance, however.

With a mixture of alcohols present, both symmetrical and unsymmetrical ethers will form. Glycols can be converted to cyclic ethers, a reaction which is favored when five-membered rings can be formed.

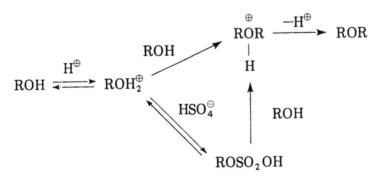

Figure 3.3 Etherification through alcohol dehydration.

In alkyd preparation, the etherification reaction is an undesirable side reaction which increases the degree of branching and enhances the risk of gelation. Since the reaction is catalyzed by both proton and Lewis acids, some of the customary esterification catalysts also accelerate the etherification.

It has been found that the acid catalyzed etherification is particularly important when a relatively strong acid, such as isophthalic acid or maleic acid, is present in the alkyd synthesis [21]. The etherification reaction is believed to be one of the reasons for the functionality of these acids apparently being greater than two (see section 3.1.4).

Etherification through alcohol dehydration may also be an alkali-catalyzed process. This means that the catalysts used to promote monoglyceride formation (see p. 79) are effective also in ether formation, as can be seen from Table 3.3 [22,23].

The etherification that occurs during alcoholysis of the triglyceride oil probably accounts for part of the differences in prop-

Table 3.3 Effect of Temperature and Catalyst (0.1%) on Ether Formation During Glycerolysis of Linseed Oil

Catalyst	Temperature (°C)	Percent oligomeric glycerol formed
Na_3PO_4	240	3.5
Na_3PO_4	280	27.3
NaOH	240	8.0
NaOH	280	25.6
CaO	240	1.3
CaO	280	7.7

Source: Ref. 22.

erties and in performance observed for alkyds that have identical chemical composition but are being prepared by the monoglyceride or the fatty acid process, respectively (see Section 3.3.2). In general, with the former method the rate of esterification slows down at somewhat higher acid values.

Transetherification, i.e., the exchange of one alkyl group for another, is very rare. It has, however, been accomplished with reactivealkyl substituents (R in Fig. 3.3), for example, diphenyl-methyl, using *p*-toluene sulphonic acid as catalyst [24].

3.2.2 Ring Formation

The polyesterification process, apart from giving linear polymers, may result in cyclic structures. The ring closure reactions are known to be of particular importance for *o*-phthalic acid-based alkyds. Structures of the type shown below have been identified in alkyd resins [25,26]:

$$
\begin{array}{c}
\text{O} \qquad\qquad\qquad \text{O} \\
\parallel \qquad\qquad\qquad \parallel \\
\text{C–O–CH}_2\text{–R–CH}_2\text{–O–C} \\
\text{C–O–CH}_2\text{–R–CH}_2\text{–O–C} \\
\parallel \qquad\qquad\qquad \parallel \\
\text{O} \qquad\qquad\qquad \text{O}
\end{array}
$$

The cyclization can be regarded as a chain-terminating reaction. The rings formed will always have fewer free functional groups than a linear chain of comparable size. Consequently, there is less chance that they will reach a high molecular weight.

The effect of ring formation on alkyd parameters, such as mean molecular weight, acid value, and hydroxyl number, has been studied by Walz [27]. An alkyd based on phthalic anhydride and pentaerythritol was fractionated by means of preparative gel permeation chromatography. The chemical compositions of the fractions were analyzed and found to be constant above a certain M_n. However, both the acid value and the hydroxyl number of the fractions increased with increasing molecular weight (see Fig. 3.4).

    The only explanation why, in spite of constant chemical composition, both the acid value and the hydroxyl number increase with increasing M_n is that ring compounds have been formed and that there are fewer cyclic structures in the higher molecular weight fractions.

    The position of the minimum on each of the curves in Figure 3.4 gives an indication of the size of the rings formed. It was estimated that the smallest rings consisted of two or three molecules each of o-phthalic acid and pentaerythritol [27].

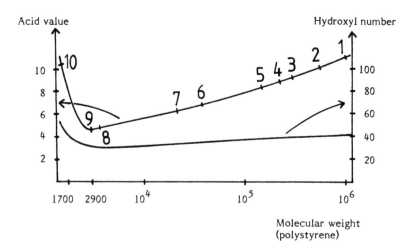

Figure 3.4 Acid value and hydroxyl number of alkyd fractions as a function of molecular weight (polystyrene used as reference). (From Ref. 27.)

The extent of the cyclization reaction may to some extent be influenced by the method of preparation. If the alkyd is prepared by the solvent-free process (the fusion process, see Section 3.3.1), the chain reaction should be favored since this is a bimolecular reaction and, thus, more dependent upon concentration than the monomolecular ring closure reaction.

In a recent work the extent of intramolecular esterification reactions was determined from data for acid value, hydroxyl number, saponification number, and M_n. Assuming ring closure to be the only reason for a higher conversion of functional groups than that corresponding to the number of polymer molecules formed by intermolecular esterification, the extent of intramolecular reactions was found to be 4-5% [54].

3.2.3 Transesterification

Two types of transesterification reactions are of relevance to alkyd synthesis, alcoholysis, and acidolysis.

Alcoholysis of esters occurs by a mechanism identical to that of ester hydrolysis—except that ROH replaces HOH—that is, by the acyl-oxygen fission mechanism. The reaction is catalyzed by both acids and bases.

$$\underset{\text{R-C-OR}^1}{\overset{O}{||}} + R^{11}OH \underset{\xleftarrow{\hspace{1.2cm}}}{\overset{H^+ \text{ or } OH^-}{\xrightarrow{\hspace{1.2cm}}}} \underset{\text{R-C-OR}^{11}}{\overset{O}{||}} + R^1OH$$

Acidolysis of esters involves alkyl-oxygen fission, i.e., the acyloxy group, $RCOO^-$, or its conjugate acid is the leaving group (the mechanism can be $S_N 1$ or $S_N 2$).

$$\underset{\text{R-CO-R}^1}{\overset{O}{||}} + R^{11}COOH \underset{\xleftarrow{\hspace{0.6cm}}}{\overset{H^+}{\xrightarrow{\hspace{0.6cm}}}} RCOOH + R^{11}\underset{}{\overset{O}{||}}COR^1$$

Normally, transesterification involving alkyl-oxygen cleavage is only important for those R^1 that give stable carbonium ions, and then only in acidic or slightly basic solutions. Typically, R^1 is tertiary alkyl, allyl, or benzyl.

It has been claimed in the literature that the polycondensation in alkyd synthesis is accompanied not only by competing alcoholysis and acidolysis but also by a direct interchange between two ester molecules [12,28]:

$$RCOOR^1 + R^{11}COOR^{111} \rightleftharpoons RCOOR^{111} + R^{11}COOR^1$$

The ester-ester interchange reaction has been postulated to involve a simultaneous making and breaking of alkoxide bonds. The mechanism involves an activated complex, the formation of which is accompanied by a large decrease in entropy. Titanium alkoxides and arsenic pentoxide have been found to catalyze the reaction.

It is unlikely, however, that a direct ester-ester interchange takes place under the conditions of alkyd synthesis. A reaction path involving solvolysis (hydrolysis, alcoholysis, or acidolysis) followed by reesterification would lead to the same end products and seems more feasible from a mechanistic point of view. The net result is still the conversion of two ester molecules into two new ones. This process has been demonstrated by isotope exchange methods for poly(hexamethylene sebacate) and deuterated diethyl succinate at 250°C [29].

3.2.4 Polyol Decomposition

Thermal dehydration of polyols may lead to end products other than ethers. Vicinal glycols are particularly sensitive and internal dehydration may lead to aldehydes, ketones, and more complicated products from subsequent addition reactions, as shown in Figure 3.5 [30-32].

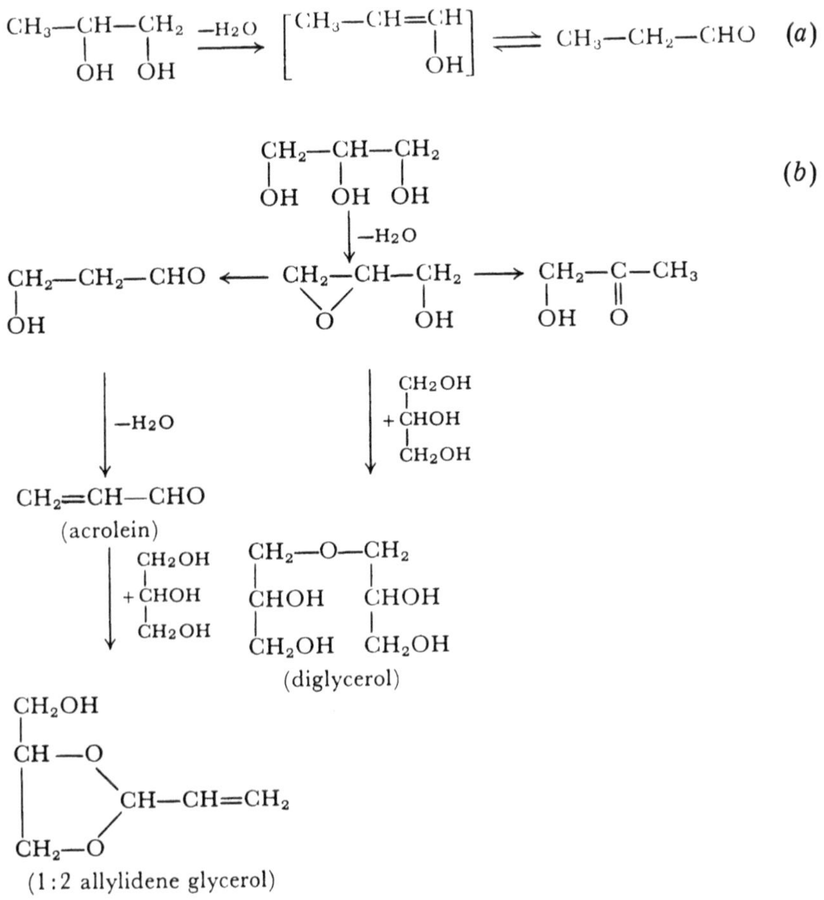

$$CH_3-\underset{\underset{OH}{|}}{CH}-\underset{\underset{OH}{|}}{CH_2} \xrightarrow{-H_2O} \left[CH_3-CH=\underset{\underset{OH}{|}}{CH} \right] \rightleftharpoons CH_3-CH_2-CHO \quad (a)$$

$$(b)$$

$$CH_2-CH-CH_2$$
$$\quad |\qquad |\qquad |$$
$$\quad OH\quad OH\quad OH$$
$$\downarrow -H_2O$$

$$CH_2-CH_2-CHO \longleftarrow CH_2-CH-CH_2 \longrightarrow CH_2-C-CH_3$$
$$|\qquad\qquad\qquad \underset{O}{\diagdown\diagup}\quad |\qquad\qquad |\quad \|$$
$$OH\qquad\qquad\qquad\qquad OH\qquad OH\quad O$$

$$\downarrow -H_2O \qquad\qquad \underset{\overset{|}{+CHOH}}{\overset{CH_2OH}{|}}$$
$$\qquad\qquad\qquad\qquad CH_2OH$$

$$CH_2=CH-CHO$$
(acrolein)

$$\underset{\overset{|}{CH_2OH}}{\overset{CH_2OH}{\underset{+CHOH}{|}}} \qquad \underset{\overset{|}{CH_2OH}}{\overset{CH_2-O-CH_2}{\underset{CHOH}{|}}} \underset{\overset{|}{CH_2OH}}{\overset{|}{CHOH}}$$
(diglycerol)

$$CH_2OH$$
$$|$$
$$CH-O$$
$$|\qquad\qquad\diagdown$$
$$\qquad\qquad CH-CH=CH_2$$
$$|\qquad\qquad\diagup$$
$$CH_2-O$$
(1:2 allylidene glycerol)

Figure 3.5 Thermal decomposition of (a) 1,2-propylene glycol and (b) glycerol.

3.2.5 Maleinization

Maleic acid, or more commonly the anhydride, is frequently employed in alkyd preparation. Apart from being a dibasic acid, it possesses a double bond capable of reacting with olefinic bonds in fatty acids from drying oils.

The reaction between maleic anhydride and unsaturated systems is well known and the process, when applied to fatty acids, is referred to by the term "maleinization" [9]. The reactions which occur during maleinization fall into one of two major categories. With conjugated systems, capable of attaining a *cisoid* configuration of the double bonds, a Diels-Alder reaction takes place at 60-80°C. With systems containing isolated double bonds, on the other hand, temperatures around 200°C are required and a complex mixture of products is usually obtained.

Early investigations showed that methyl oleate reacts with 1 mole, methyl linoleate with 2 moles and methyl linolenate with 2.5 moles of maleic anhydride at 200-210°C [33]. Analysis indicated that the degree of unsaturation of the fatty acids was not affected by the reaction. The simplest maleinization, that of oleic acid, has been widely studied. Two different routes leading to different reaction products have been proposed. In the first of these, the reaction occurs at the allylic positions in the fatty acid, leading to reaction products with the unsaturation retained at the 9-10 position (Fig. 3.6, path *a*) [33,34]. By the alternative route, an ene reaction takes place, giving rise to products where the double bond of the fatty acid has undergone an allylic shift (Fig. 3.6, path *b*) [35,36].

More recent investigations seem to show that both types of reaction take place simultaneously [37,38]. In one investigation ethyl oleate (ethyl *cis*-9-octadecenoate) was treated with maleic anhydride at 210°C. The anhydride ring was opened with alkali and the reaction product was oxidized in an aqueous solution of periodate containing catalytic amounts of permanganate (Fig. 3.7) [37]. Under these conditions, the olefins are readily cleaved to carboxylic acids.

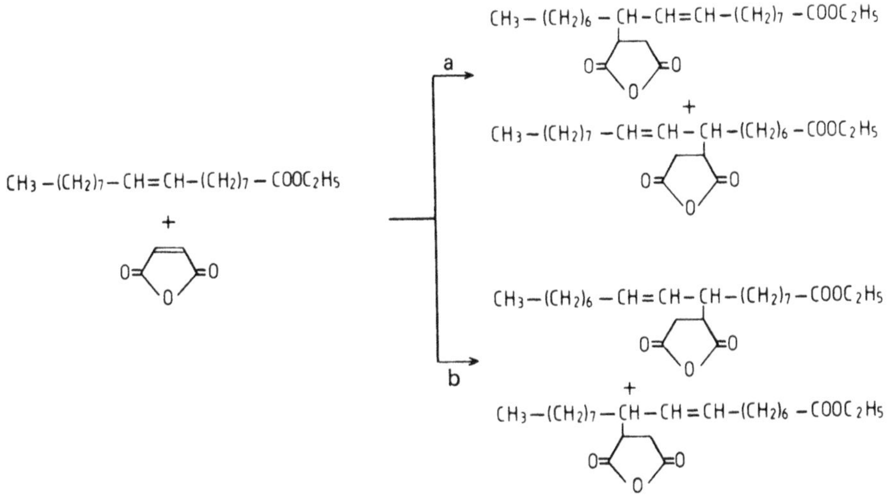
CH3-(CH2)6-CH-CH=CH-(CH2)7-COOC2H5
a
+
CH3-(CH2)7-CH=CH-CH-(CH2)6-COOC2H5
CH3-(CH2)7-CH=CH-(CH2)7-COOC2H5
+
CH3-(CH2)6-CH=CH-CH-(CH2)7-COOC2H5
b
+
CH3-(CH2)7-CH-CH=CH-(CH2)6-COOC2H5

Figure 3.6 Maleinization of ethyl 9-octadecenoate proceeding by two different routes.

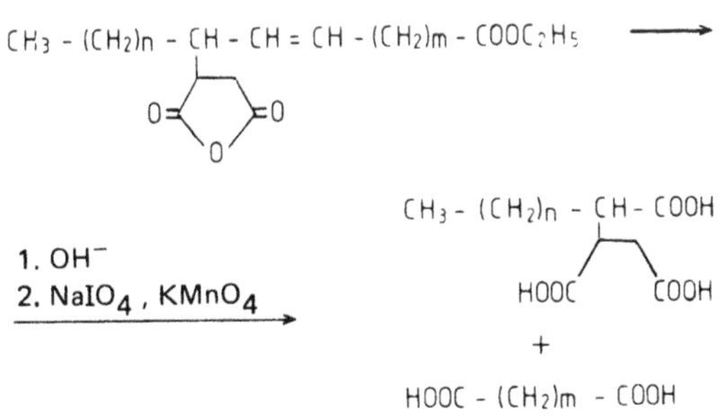
CH3-(CH2)n-CH-CH=CH-(CH2)m-COOC2H5
1. OH⁻
2. NaIO4, KMnO4
CH3-(CH2)n-CH-COOH
HOOC COOH
+
HOOC-(CH2)m-COOH

Figure 3.7 Oxidative cleavage of the addition product of maleic anhydride and ethyl 9-octadecenoate.

If the addition of maleic anhydride proceeds according to reaction *a* of Figure 3.6, the reaction sequence of Figure 3.7 would give nonanoic acid and nonanedioic acid, as well as succinic acid derivatives from the substituted part of the ethyl oleate. If, on the other hand, the addition follows path *b* of Figure 3.6, a mixture of octanoic acid and octanedioic acid would be obtained along with succinic acid derivatives.

Analysis by GLC shows an approximate 1:2 ratio of cleavage products according to reactions *a* and *b*, as can be seen from Table 3.4. Consequently, both the stepwise reaction, having a free radical chain mechanism, and the ene reaction seem to be of importance in this case. Interestingly, the ethyl ester of elaidic acid, the *trans* isomer of 9-octadecenoic acid, in a similar experiment gave fission products almost exclusively corresponding to the ene reaction (Table 3.4).

Addition of maleic anhydride to the more highly unsaturated fatty acid esters have not been investigated in full detail. From a quantitative point of view, however, maleinization is likely

Table 3.4 Relative Amounts of C-8 and C-9 Acids after Maleinization of Ethyl 9-Octadecenoates and Subsequent Oxidative Cleavage

	Fission products (%)			
Starting products	Octanoic acid	Octanedioic acid	Nonanoic acid	Nonanedioic acid
Ethyl oleate	28	36	12	24
Ethyl elaidate	46	41	5	8
Ethyl oleate with radical inhibitor	41	42	6	11

Source: Ref. 37.

to be even more important with these compounds. In the preparation of drying oil alkyds the actual functionality of maleic anhydride is, therefore, considerably greater than 2 (see also p. 68).

Another side reaction involving maleic acid is the addition of a polyol to the olefinic bond. This reaction, although far less important than the maleinization, should be taken into consideration, particularly in the preparation of nondrying alkyds. The reaction has special relevance to the preparation of unsaturated polyesters and will be further discussed in Section 6.5.2.

3.3 MANUFACTURING PROCESS

Alkyds are prepared by polycondensation of the acid and alcohol components until a predetermined acid value-viscosity relationship has been achieved. The reaction is normally performed under inert gas or solvent vapor to minimize oxidation of unsaturated components. In the initial phases of the reaction, the drop in the acid value is rapid and the increase in the viscosity is slow. Towards the later stages of the reaction, the reverse is true. The preparation may be performed by a solvent-free process, the fusion method, or by using a small amount of a solvent which forms an azeotrope with water, the solvent method. Furthermore, in the preparation of fatty acid modified polyesters, either the triglyceride or the fatty acid derived from it may be employed as starting materials. These two procedures are referred to as the monoglyceride and the fatty acid process, respectively.

3.3.1 Fusion Method vs. Solvent Method

The fusion method is the older process but still widely used, especially for alkyds of an oil length of 60% and more. The

reaction is carried out at a temperature of 220-250°C, and the
inert gas sparge, which is used for dewatering, also causes some
loss of volatile polyols and of phthalic anhydride. Thus, the
fusion method brings about considerable problems associated with
loss by volatilization, and it is not the method of choice for preparing alkyds with narrow specification limits [39].

The solvent method (sometimes called the azeotropic
method) is preferred for high solids alkyds where the resin composition is critical. In this process the esterification is performed in
the presence of a small quantity of water-immiscible solvent,
usually xylene. The process is carried out under continuous
azeotropic distillation of the solvent. The xylene-water vapor
mixture is condensed, the water is separated and the organic dis-
tillate is returned to the reactor. The reaction temperature is
governed by the refluxing temperature which, in turn, depends on
the amount of xylene used, 5% being a normal value.

The solvent method offers much better control of the
resin composition, as there is virtually no loss of polyols or other
low molecular materials by volatilization, and the amount of
phthalic anhydride lost by sublimation is also markedly reduced.
Furthermore, the reduction of viscosity of the reaction mixture
and the improved control of heat transfer minimizes local over-
heating and enables a more uniform alkyd, free from gel particles,
to be prepared. It has been claimed that the presence of an inert
solvent changes the gel point and enables the preparation of resins
of higher molecular weight than can be produced by the fusion
process [40]. This is of particular relevance to the preparation of
high solids alkyd resins, as these are often formulated to give a
lower molecular weight than is desired from a film performance
point of view.

3.3.2 Monoglyceride vs. Fatty Acid Process

Since triglycerides, such as soya bean oil and safflower oil, are the ultimate source of fatty acids, it seems obvious to use these as starting materials in the production of alkyds and rely on transesterification reactions to get an even distribution of the fatty acids along the polyester backbone. However, when a triglyceride oil is heated together with polyols and dibasic acids, the polyols react preferentially with the acids and a heterogeneous mixture of triglyceride and unmodified polyester is obtained. The way to overcome this problem is to perform a controlled transesterification of the fatty acids prior to the condensation step. This is usually done by reacting one mole of triglyceride with two moles of glycerol (or the equivalent amount of another polyol) at a temperature of 220-250°C until the monoglyceride stage is attained. The alcoholysis reaction, when performed with glycerol, is assumed to give monoester with the acyl group in 1-position. In reality, of course, the reaction product is a complex mixture of free glycerol, mono-, and diglycerides. The preferred catalysts are PbO, $Ca(OH)_2$, and Ca soaps [41].

$$
\begin{array}{ccccc}
CH_2OCOR & & CH_2OH & & CH_2OCOR \\
| & & | & \xrightarrow[\text{catalyst}]{\text{alkaline}}^{\Delta} & | \\
CHOCOR & + & 2\ CHOH & & 3CHOH \\
| & & | & & | \\
CH_2OCOR & & CH_2OH & & CH_2OH
\end{array}
$$

Alternatively, acidolysis, i.e., reacting the triglyceride with a dibasic acid, may be performed to overcome the problem of incompatibility. This process results in triesters of glycerol having one or two fatty acid residues replaced by monoesters of the dibasic acid, in addition to free fatty acid. The reaction is more sluggish than the alcoholysis and is not widely used, however.

The main advantage of acidolysis over alcoholysis is that it can reduce the solubility problems sometimes encountered with iso- and terephthalic acid when added to the monoglyceride mixture.

After formation of monoglyceride the di- and tribasic acids and the rest of the polyols are added and the condensation is carried out until the desired viscosity-acid value relationship is reached.

In the fatty acid process the triglyceride is not used, but the fatty acid itself. Apart from giving a greater freedom in the choice of polyol components, this process is more reproducible and gives better control over molecular weight and molecular weight distribution of the resin [9]. This is particularly important in the preparation of high solids alkyd resins, and the fatty acid process is the most popular method for this purpose.

Comparison of the two processes has shown that with the fatty acid process the condensation can be carried on to lower acid values, which is advantageous for the drying properties. In addition, only the fatty acid process permits incremental addition of fatty acid (and other raw materials) during the course of the reaction. Incremental addition of raw materials is of particular interest in the preparation of high solids alkyds, as described in the next section.

It should be pointed out that the choice of manufacturing process is not only a matter of finding the most suitable synthesis procedure. There exist distinct differences in film properties between alkyds of the same composition but prepared by different processes [42,43]. Evidently, there are considerable diversities in polymer structure depending on the method of preparation. For instance, it has been claimed that the fatty acid residues are situated mainly in the 1-position of the glycerol molecule in resins prepared by the monoglyceride process and in the 2-position in alkyds made by the fatty acid process [44]. This has been used as evidence that interchange reactions are slow compared to polyesterification [49]. (See also p. 68 regarding differences in chemical structure due to etherification reactions.)

3.4 STEP-GROWTH METHOD

One way of synthesizing alkyds of narrow molecular weight distribution is to use reactive derivatives of acids and alcohols as starting materials. The reaction temperature can then be kept low, which minimizes the side reactions discussed in Section 3.2. The commonly used o-phthalic anhydride is, of course, one example of a reactive derivative, but in order to make use of the concept, at least one more reagent capable of reacting at a temperature well below normal esterification temperature is needed.

In a model experiment the alkyd shown in Figure 3.8 has been prepared, and in this synthesis the reactivity of anhydrides and oxirane rings has alternately been exploited to keep the esterification temperature down [45]. In the final steps, which involve reaction between a primary alcohol and a carboxylic acid, various esterification catalysts have been tested. The reaction sequences appear in Figure 3.9.

The alkyd was built up stepwise, starting from adipic acid. One mole of adipic acid was reacted with two moles of Cardura at 130-140°C. The oxirane ring was then opened, but the temperature was too low for the resulting secondary hydroxyl group to react further.

Two moles of phthalic anhydride was added to the resulting glycol, the temperature being kept so low (140-150°C) that the aromatic carboxyl group formed remained intact. Two moles of Cardura and two moles of phthalic anhydride were then added

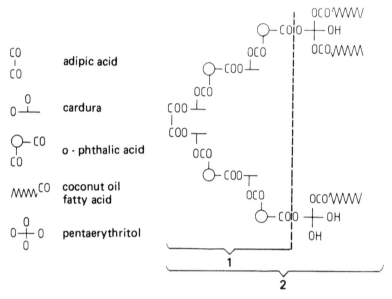

Figure 3.8 Alkyd prepared by the step-growth method.

stepwise, producing dicarboxylic acid *1* (*1* is the alkyd part to the left of the dashed line in Figure 3.8).

Different routes (see Fig. 3.9) were tested for the last two steps of the alkyd synthesis:

A: Two moles of pentaerythritol were reacted with *1*. After completed reaction, 3 moles of coconut fatty acid were added and the heating was continued. No catalyst was used and the reaction temperature was kept at 190 and 200°C, respectively.

B: The same reactions as above were performed but in the presence of dicyclohexylcarbodiimide (DCC) and pyridine and with *p*-toluene sulphonic acid (pTSA) as catalyst. The reactions proceeded rapidly at room temperature.

C: Two moles of pentaerythritol were reacted with three moles of coconut oil fatty acid at room temperature in the presence

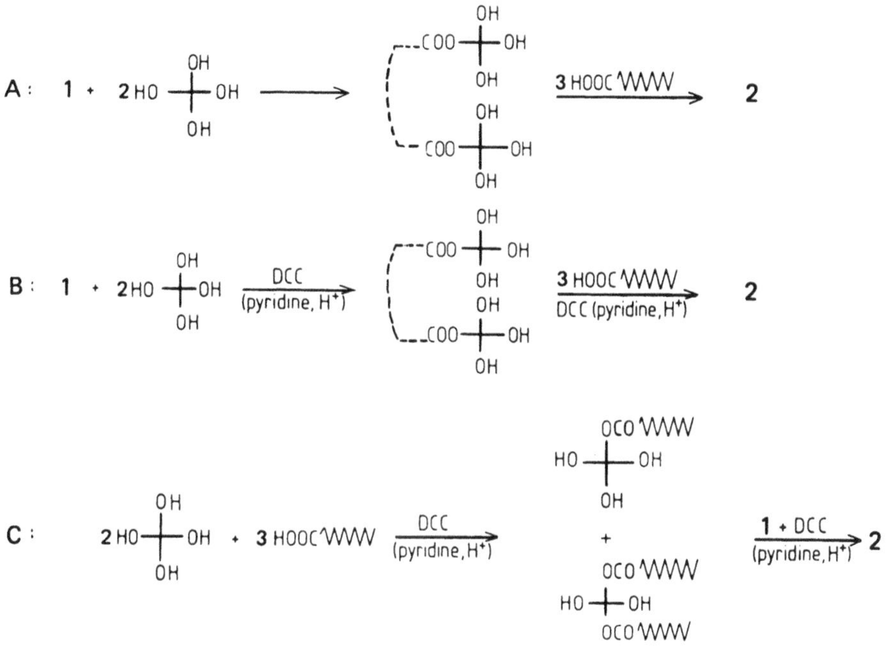

Figure 3.9 Reaction sequences for the synthesis of the alkyd of Figure 3.8.

of DCC and pyridine-pTSA. The resulting product was then reacted with *1* with the aid of the same reagents.

The alkyd structure according to Figure 3.8 and the reaction sequences of Figure 3.9 are, of course, idealized. The reactions were run in a yield of at least 95% calculated from volumetric determinations. The kinetics of the reactions, which involve ring opening, was also studied with ^{13}C-NMR. The oxirane carbon atoms, as well as the carbonyl carbons in the anhydride ring, have characteristic chemical shifts and can be followed down to low concentrations even in these complex systems. To minimize side reactions, it is important not to allow the reaction to proceed longer than necessary.

A long series of esterification catalysts were tested for the two last steps. Out of the strong acids tested, methane sulphonic acid had the best effect. This acid, however, proved to catalyze the etherification reaction also, resulting in a rapid increase of the viscosity already at relatively high acid values.

The use of the system DCC-pyridine-pTSA proved to be the most effective and most selective esterification method. The reactions were completed in less than one hour at room temperature. A relatively good selectivity was also obtained without the strong acid. DCC was added in equimolar amounts and functions as a dehydration agent under formation of dicyclohexyl urea.

In all cases, the alkyd was run to an acid value of around 20. The oil length of the alkyd was 31% and the theoretical hydroxyl number 90 mg KOH/g. The three types, which from a common intermediate (1) were prepared according to routes A, B, and C, were analyzed with regard to molecular weight and molecular weight distribution by GPC.

The GPC curves are shown in Figure 3.10. Route A, which illustrate the normal process in alkyd synthesis, gives a considerably broader molecular weight distribution than routes B and C. Also, the average molecular weight of the product according to route A is considerably higher than that of the products according to the other routes.

It further appears from Figure 3.10 that route C gave a more defined product than route B. This is probably due to the dicarboxylic acid 1 having a greater tendency to react with more than one hydroxyl group in pentaerythritol in route B than in route C, since in the latter route 1-2 of the hydroxyl groups of the polyol are already esterified.

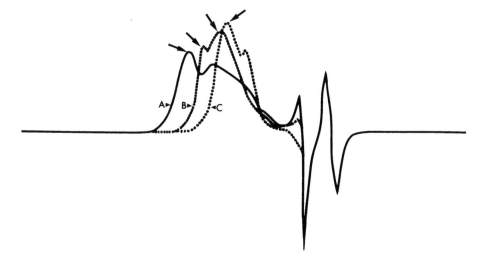

Figure 3.10 Gel permeation chromatography (GPC) curves of alkyds A, B, and C of Figure 3.9. The arrows indicate, from left to right, molecular weights 50,000, 30,000, 8,000, and 4,000.

The viscosity as a function of the solids content of solutions of the three alkyds in ethyl glycol and in xylene is shown in Figure 3.11. Alkyd A, which has the highest molecular weight and the broadest molecular weight distribution, has, as expected, the highest viscosity in both solvents. It can also be seen that there are substantial viscosity differences between the alkyds prepared by route B and route C, which well correspond to the GPC analyses of these alkyds.

In a test of film properties (Table 3.5) it was found that alkyd A gave the hardest and alkyd C the softest film. This behavior is probably due to the alkyds having different amounts of low molecular fractions, which function to some extent as a plasticizer.

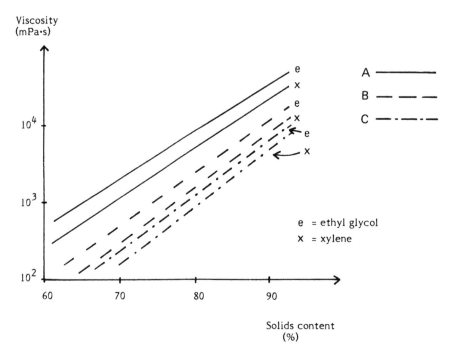

Figure 3.11 Viscosity as a function of solids content for alkyds A, B, and C of Figure 3.9.

Table 3.5 Hardness and Elasticity of Films Made from Alkyds Prepared by Routes A, B, and C of Figure 3.9 in Combination with CIBAMIN M 100.

Alkyd	Hardness (mPa)	Elasticity (%)
A	98	40
B	94	62
C	80	70

Other reactive derivatives that are sometimes used to attain low temperature polyesterifications are the anhydrides of maleic acid and trimellitic acid. It seems probable that a striving towards alkyds of high solids content entails an increased interest for this type of raw material. Acid chlorides and epoxidized compounds are examples of substances that may become of current interest for this purpose.

The price of the raw materials will, of course, always be a restricting factor. DCC, used in the above example to attain an extremely selective esterification, is hardly economically realistic today even though the product can be regenerated. Trifluoroacetic anhydride is another reagent that gives a high degree of selectivity in alkyd synthesis. Even though this substance, unlike DCC, need not be used in equimolar amounts, it is at present too expensive to be used in normal cases. It is, however, possible that in the future higher raw material costs will be accepted if alkyd paints of substantially higher solids content can thereby be provided.

It is likely that the syntheses must be performed stepwise to a greater extent than before in order to obtain well-defined products. It is also possible that a better result will be attained if part of the alkyd is prepared separately in order to be subsequently reacted with a prepolymer, as illustrated by route C in Figure 3.9. This, of course, also makes the process more expensive.

An alternate way of preparing alkyds of relatively well-defined structure is to defunctionalize the multifunctional polyol with a monobasic acid prior to the polyesterification step [46]. In the example given, the monofunctional acid is pelargonic acid and the ratio of this component to the polyol is chosen so that, on the average, a difunctional alcohol is obtained. This compound is then further reacted with dicarboxylic acids to produce a linear polyester. Hydroxyl-terminated resins suitable for high solids alkyd-melamine resin combinations were obtained by this approach.

With the reverse order of addition, i.e., an incremental addition of the monobasic acid to a partly polymerized polyol/dibasic acid mixture, a resin of higher viscosity is obtained [40]. This method, which has been referred to as the high polymer technique, has been used as a means of preparing alkyds having improved film hardness and drying characteristics [47]. The high polymer technique may not be applicable to the synthesis of high solids alkyds, but it is a further example of the importance of the order and time of addition of the ingredients.

3.5 BLOCKING OF POLAR GROUPS

As shown in Section 2.4, polar groups, in particular hydroxyl groups, to a large extent govern the viscosity of alkyd resins. The hydroxyl groups are the main hydrogen bond donors of the polymers, while ester groups function as acceptors. Thus, a low concentration of hydroxyl groups in the resin minimizes the intermolecular forces and gives an alkyd of low viscosity.

Hydroxyl groups, however, are needed in the resins for various reasons. Baking alkyds require a certain hydroxyl number for adequate curing with amino resins. Air-drying resins need a minimum concentration of hydroxyl groups to give proper adhesion and pigment wetting. Consequently, a mere reduction in the hydroxyl number of the alkyd is not a fruitful way to attain high solids resins.

If, however, the hydroxyl groups were reacted with a blocking agent which spontaneously was removed during curing,

the concept of low hydroxyl number could be used to raise the
solids content of alkyd resins. Recently, this approach has been
tested using acetic anhydride, dihydropyran, and chlorotrimethylsilane as blocking agents [48]. The blocking reactions are shown
in Figure 3.12.

Acetic anhydride readily reacts with hydroxyl groups to
form acetates. Dihydropyran is a widely used OH-protecting
agent, and the blocking procedure involves formation of the tetrahydropyranyl ether. Chlorotrimethylsilane, in the presence of an
alkaline catalyst, reacts with both hydroxyl and carboxyl groups
forming trimethylsilyl ethers and esters, respectively.

The tetrahydropyranyl and the trimethylsilyl blocking
groups are known to be extremely acid sensitive and would be
expected to be easily removed under the acid-catalyzed curing
with an amino resin [49]. The acetyl group would not be expected to be as easily deblocked under these conditions. However,
since p-toluene sulphonic acid has been shown to catalyze the
hydrolysis of normal ester bonds in alkyds during curing [50],
deacetylation should proceed at a reasonable rate, provided a
sufficient amount of acid catalyst was used.

The three blocking procedures shown in Figure 3.12 were
applied to two alkyds having hydroxyl numbers 86 and 110 and
acid values of 24 and 10, respectively. A decrease in viscosity
corresponding to a raise in solids content of 5-10% was obtained
with acetic anhydride and chlorotrimethylsilane as blocking agents.
Blocking with dihydropyran gave no marked effect on the viscosity, however.

The blocked alkyds were cured with a melamine resin of
HMMM type (see p. 145) using p-toluene sulphonic acid as catalyst.
The tetrahydropyranyl and the trimethylsilyl derivatives behaved

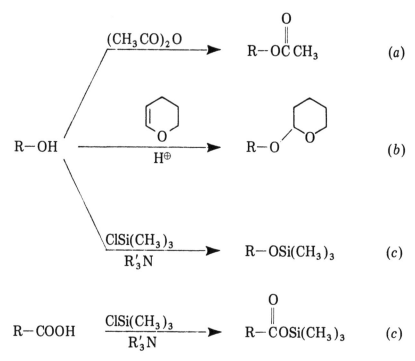

Figure 3.12 Blocking of polar groups with (a) acetic anhydride, (b) dihydropyran, and (c) chlortrimethylsilane.

in the normal way, giving essentially the same film characteristics as the unblocked alkyds. The acetylated alkyds, on the other hand, gave insufficient curing and soft, tacky films. Evidently, under the conditions used, the deblocking is sufficiently facile with the two former derivatives but not with the latter.

Consequently, out of the three blocking procedures evaluated only the silylation (path c of Fig. 3.12) was found to combine a good viscosity lowering effect with acceptable (retained) curing characteristics. For practical purposes, however, silanes are too expensive as raw materials. The work should be regarded as model experiments but, nevertheless, the results imply that the approach may be a fruitful one in the search for methods to raise the solids content of alkyd resins. Inexpensive OH-protecting groups, easily removed under acidic conditions, such as di- and trichloroacetyl, could be candidates for the future.

3.6 COLLOIDAL ALKYDS

It has been suggested that gelation of an alkyd is preceded by the formation of microgel particles in the form of an organosol. The particles become more and more crowded and eventually coalesce into a macrogel [40].

Alkyds having a substantial fraction of microgel particles may be referred to as colloidal alkyds. They show a remarkable viscosity-solids content relationship since the high molecular fraction, being in a colloidal state, contributes only little to the resin viscosity. A series of air-drying alkyd resins were prepared, all having an oil length of 52% and a mole ratio of fatty acid (linoleic acid):glycerol:phthalic anhydride of 1.0:1.36:1.11. As starting materials varying mixtures of previously synthesized α-monoglyceride, α,α'-diglyceride, and triglyceride of linoleic acid were employed [51].

All alkyds were processed to the same acid value, i.e., to the same number average molecular weight. The measured molecular weight distribution of the resins differed, the alkyd prepared from α-monoglyceride having the highest amount of high molecular weight fraction. This resin also gave the highest solids content in white spirit at a given viscosity. This behavior seems contradictory to the discussion of the role of the molecular weight fractions, given in Section 2.2.2. However, it has been explained by the assumption that the high molecular material is not present as a solute but as a microgel.

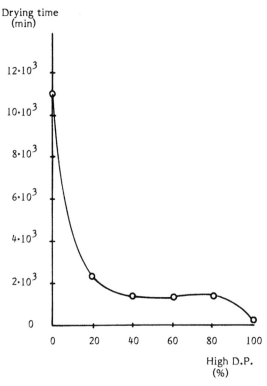

Figure 3.13 Influence of high molecular weight fraction on dry-ing time of an alkyd. (From Ref. 40.)

Preparation of alkyds having a high fraction of microgel may be used as a means of achieving high solids resins. Also, the rate of drying is markedly reduced when microgel fractions are present [51]. The importance of particles of microgel for fast drying has been demonstrated by mixing a high and a low molecular weight alkyd. As is shown in Figure 3.13, as little as 20% high molecular weight fraction gives an enormous reduction of the drying time for a linoleic-based alkyd [40].

It has also been found that alkyds possessing a substantial fraction of microgel may give low gloss of films when cross-linked with a melamine-formaldehyde resin [51]. This is probably due to the hydroxyl groups in the organosol not being fully available for pigment wetting. Determination of hydroxyl number by acetylation also gave a very low value compared with the theoretical number.

Alkyds having a substantial portion of colloidal fraction (defined as having molecular weight above 10^5) have been prepared by using a carefully designed heating schedule in the preparation [52,53]. Although mixing problems appeared with certain melamine-formaldehyde resins, film properties obtained with the colloidal alkyds were generally superior to those obtained with alkyds of the same formulation but processed in a conventional way.

3.7 PHYSICAL PROPERTIES OF ALKYD RAW MATERIALS

Physical properties of alkyd raw materials are given in Table 3.6-3.11, pp. 94-101.

Table 3.6 Physical Constants of Glycols

Glycol	Formula	Molecular weight	Melting point °C	Boiling range °C	Specific gravity 20/20°C
Ethylene glycol	$HO(CH_2)_2OH$	62.07	−13	193-205	1.12
Propylene glycol	$CH_3CH(OH)CH_2OH$	76.1	−60[a]	185-189	1.04
1,3-Butylene glycol	$CH_3CH(OH)CH_2CH_2OH$	90.2	23-27	207.5	1.01
Neopentyl glycol	$HOCH_2C(CH_3)_2CH_2OH$	104.15	120-130	213	1.07, 25/4°
Diethylene glycol	$HO(CH_2)_2O(CH_2)_2OH$	106.12	−8	242-250	1.12
Dipropylene glycol	$CH_3CHOHCH_2OCH_2CHOHCH_3$	134.18	−40[b]	228-236	1.02

Name	Structure	Mol wt	mp/Pour point	bp	Density
1,4-Cyclohexanedimethanol	HOCH$_2$(C$_6$H$_{10}$)CH$_2$OH	144.21	35	285	1.15
2,2,4-Trimethyl-1,3-pentanediol	$\overset{\text{OH}}{|}$ (CH$_3$)$_2$CHCHC(CH$_3$)$_2$CH$_2$OH	146.22	46-55	225	0.93, 55/15°
Triethylene glycol	HO(CH$_2$CH$_2$O)$_3$H	150.17	-7.2	278-300	1.13
2,2-Dimethyl-3-hydroxypropyl-2,2-dimethyl-3-hydroxypropionate	$\overset{\text{OH}}{|}$ $\quad$ $\overset{\text{O}}{\underset{||}{}}$ $\quad$ $\overset{\text{OH}}{|}$ CH$_2$C(CH$_3$)$_2$CH$_2$OCC(CH$_3$)$_2$CH$_2$	204.26	46-51	292.5	1.0234, 50/20°
Bisphenol A	(CH$_3$)$_2$C(C$_6$H$_4$OH)$_2$	228.3	154.5	—	1.195, 25/25°
Hydrogenated Bisphenol A	(CH$_3$)$_2$C(C$_6$H$_{10}$OH)$_2$	240.35	150	—	—

[a]Sets to glass below this temperature.
[b]Pour point.
Source: Ref. 55.

Table 3.7 Physical Constants of Higher Polyols

Polyol	Formula	Molecular weight	Melting point (°C)	Boiling point (°C)	Specific gravity (20/20°C)
Glycerol	$\begin{matrix} & OH \\ & | \\ HOCH_2CHCH_2OH \end{matrix}$	92.1	18	290	1.26
Trimethylolethane	$CH_3C(CH_2OH)_3$	120.15	185–195	—	1.22
Trimethylolpropane	$CH_3CH_2C(CH_2OH)_3$	134.1	60	295	1.14
Pentaerythritol (Mono)[a]	$C(CH_2OH)_4$	136.15	250–258	—	1.395, 25/4°C
Pentaerythritol (Tech)[b]	$C(CH_2OH)_4$	136.15 254.3	180–240	—	1.38, 25/4°C
Dimethylolpropionic acid	$CH_3C(CH_2OH)_2COOH$	134	178–180	—	1.355, 25°C

[a]98.5% mono.
[b]88% mono, 12% di.
Source: Ref. 55.

Table 3.8 Physical Constants of Monofunctional Acids

Acid	Formula	Molecular weight	Melting point (°C)	Boiling point (°C)	Specific gravity (20/20°C)
Crotonic	$CH_3 CH = CHCOOH$	86.1	72	182	0.97
Caproic	$CH_3(CH_2)_4 COOH$	116.2	—	206	0.93
Benzoic	$C_6 H_5 COOH$	122.1	122	249	1.27
2-Ethylhexoic	$C_4 H_9 CH(C_2 H_5)COOH$	144.2	−118	227	0.908
Caprylic	$CH_3(CH_2)_6 COOH$	144.2	17	238	0.91
Pelargonic	$CH_3(CH_2)_7 COOH$	158.2	11	253	0.91
Capric	$CH_3(CH_2)_8 COOH$	172.3	32	270	0.89
p-tert-Butylbenzoic	$(CH_3)_3 C(C_6 H_4)COOH$	178.1	165	—	1.15

Source: Ref. 55.

PREPARATION OF HIGH SOLIDS ALKYDS

Table 3.9 Physical Constants of Polyfunctional Acids and Anhydrides

Acid or anhydride	Formula	Molecular weight	Melting point (°C)	Boiling point (°C)	Specific gravity (20/20°C)
Maleic anhydride	$CHC\diagup\!\!\!{}^{O}$ $\|\quad\!\!{}^{O}$ $CHC\diagdown\!\!{}_{O}$	98.1	53	200	0.934 20/4°
Fumaric acid	$CHCOOH$ $\|$ $CHCOOH$	116.1	200 (sublimes)	—	1.63
Succinic acid	$(CH_2)_2(COOH)_2$	118	185	235	1.55
Adipic acid	$(CH_2)_4(COOH)_2$	146	152	—	1.37
Phthalic anhydride (ortho)	$C_6H_4(CO)_2O$	148.2	131	285	1.53[a], 4°C
Tetrahydrophthalic anhydride	$C_6H_8(CO)_2O$	152.1	99	—	1.2[a], 105°C
Hexahydrophthalic anhydride	$C_6H_{10}(CO)_2O$	154.2	35	—	1.18[a], 40°C
Isophthalic acid (meta)	$(C_6H_4)(COOH)_2$	166.2	345-348	—	1.54
Terephthalic acid (para)	$(C_6H_4)(COOH)_2$	166.2	300 (sublimes)	—	—
Trimellitic anhydride	$HOOC(C_6H_3)(CO)_2O$	192	165	—	1.55
Azelaic acid	$(CH_2)_7(COOH)_2$	188.2	106	286.5, 100 mm	1.03
Sebacic acid	$(CH_2)_8(COOH)_2$	202	133	295, 100 mm	1.11
Pyromellitic dianhydride	$O(CO)_2C_6H_2(CO)_2O$	218	286	400	1.68
Tetrachlorophthalic anhydride	$C_6Cl_4(CO)_2O$	285.9	254-255	—	1.49[a], 275°C
Tetrabromophthalic anhydride	$C_6Br_4(CO)_2O$	463.7	270	—	—

[a]Density in g/ml.
Source: Ref. 55.

Table 3.10 Structure and Physical Properties of Principal Fatty Acids in Oils

Fatty acid	Scientific name	Structure[a]	Empirical formula	Mol. wt.	MP (°C)	BP (°C)[b]	Density (20/4°)[c]
Arachidonic	5,8,11,14-Eicosatetraenoic	$CH_3 \cdot (CH_2)_4 \cdot CH{:}CH \cdot CH_2 \cdot CH{:}CH \cdot CH_2 \cdot CH{:}CH \cdot CH_2 \cdot CH{:}CH \cdot CH_2 \cdot CH(CH_2)_3 \cdot COOH$	$C_{20}H_{32}O_2$	304.5	−80	d	0.9356
Clupanodonic	4,8,12,15,19-Docosapentaenoic	$CH_3 \cdot CH_2 \cdot CH{:}CH \cdot (CH_2)_2 \cdot CH{:}CH \cdot CH_2 \cdot CH{:}CH[(CH_2)_2 \cdot CH{:}CH]_2 \cdot (CH_2)_2 \cdot COOH$	$C_{22}H_{34}O_2$	330.6	−78	326[5]	
Eleostearic	9,11,13-Octadecatrienoic						
Alpha	*cis-trans-trans*	$CH_3 \cdot (CH_2)_3 \cdot CH{:}CH \cdot CH{:}CH \cdot CH{:}CH(CH_2)_7 \cdot COOH$	$C_{18}H_{30}O_2$	278.5	49	235[3]	0.8965
Beta	*trans-trans-trans*				71		0.8839**
Licanic	4-Keto-9,11,13-octadecatrienoic	$CH_3 \cdot (CH_2)_3 \cdot CH{:}CH \cdot CH{:}CH \cdot CH{:}CH(CH_3)_4 \cdot \overset{O}{\overset{\|}{C}}(CH_2)_2 \cdot COOH$	$C_{18}H_{28}O_3$	292.5			
Alpha	*cis-trans-trans*				74.7		
Beta	*trans-trans-trans*				99.5		
Linoleic	9,12-Octadecadienoic	$CH_3 \cdot (CH_2)_4 \cdot CH{:}CH \cdot CH_2 \cdot CH{:}CH \cdot (CH_2)_7 \cdot COOH$	$C_{18}H_{32}O_2$	280.5	−5.0	229–230[16]	0.9025
	10,12-Octadecadienoic	$CH_3 \cdot (CH_2)_4 \cdot CH{:}CH \cdot CH{:}CH \cdot (CH_2)_8 \cdot COOH$			56–57		0.8686*
Linolenic	9,12,15-Octadecatrienoic	$CH_3 \cdot CH_2 \cdot CH{:}CH \cdot CH_2 \cdot CH{:}CH \cdot CH_2 \cdot CH{:}CH(CH_2)_7 \cdot COOH$	$C_{18}H_{30}O_2$	278.5	−11	230–232[17]	0.9157
Myristic	Tetradecanoic	$CH_3 \cdot (CH_2)_{12} \cdot COOH$	$C_{14}H_{28}O_2$	228.4	54.3	149.3[1]	0.8439**
Oleic	9-Octadecenoic	$CH_3 \cdot (CH_2)_7 \cdot CH{:}CH \cdot (CH_2)_7 \cdot COOH$	$C_{18}H_{34}O_2$	282.5	13.4	286[100]	0.894
Palmitic	Hexadecanoic	$CH_3 \cdot (CH_2)_{14} \cdot COOH$	$C_{16}H_{32}O_2$	256.5	63	167[1]	0.8474**
Palmitoleic	9-Hexadecenoic	$CH_3 \cdot (CH_2)_5 \cdot CH{:}CH(CH_2)_7 \cdot COOH$	$C_{16}H_{30}O_2$	254.5	0.5–10.5		
Stearic	Octadecanoic	$CH_3 \cdot (CH_3)_{16} \cdot COOH$	$C_{18}H_{36}O_2$	284.6	70.1	183.5[1]	0.9408

[a] All ethylene groups are *cis* unless otherwise indicated.
[b] Superscript denotes mm pressure.
[c] * = 70; ** = 80.
[d] Decomposes.
Source: Ref. 56.

Table 3.11 Typical Fatty Acid Composition of Vegetable Oils (in %)

	Fatty Acids		Castor	Chinese Tallow	Coconut	Corn
C_4	Butyric (*Butanoic*)	$C_4H_8O_2$				
C_6	Caproic (*Hexanoic*)	$C_6H_{12}O_2$			x	
C_8	Caprylic (*Octanoic*)	$C_8H_{16}O_2$			6	
C_{10}	Capric (*Decanoic*)	$C_{10}H_{20}O_2$			6	
C_{12}	Lauric (*Dodecanoic*)	$C_{12}H_{24}O_2$		2	44	
	Lauroleic (*cis-9-Dodecenoic*)	$C_{12}H_{22}O_2$				
C_{14}	Myristic (*Tetradecanoic*)	$C_{14}H_{28}O_2$		4	18	
	Myristoleic (*cis-9-Tetradecenoic*)	$C_{14}H_{26}O_2$				
C_{16}	Palmitic (*Hexadecanoic*)	$C_{16}H_{32}O_2$	2	66	11	13
	Palmitoleic (*cis-9-Hexadecenoic*)	$C_{16}H_{30}O_2$				
C_{18}	Stearic (*Octadecanoic*)	$C_{18}H_{36}O_2$	1	1	6	4
	Oleic (*cis-9-Octadecenoic*)	$C_{18}H_{34}O_2$	7	27	7	29
	Ricinoleic (*12-Hydroxy-cis9-Octadecenoic*)	$C_{18}H_{34}O_3$	87			
	Linoleic (*cis-9, cis-12-Octadecadienoic*)	$C_{18}H_{32}O_2$	3		2	54
	Linolenic (*cis-9, cis-12, cis-15-Octadecatrienoic*)	$C_{18}H_{30}O_2$			x	
	Eleostearic (*cis-9, trans-11, trans-13-Octadecatrienoic (?)*)	$C_{18}H_{30}O_2$				
	Licanic (*4-Keto-9, 11, 13-Octadecatrienoic*)	$C_{18}H_{28}O_3$				
C_{20}	Arachidic (*Eicosanoic*)	$C_{20}H_{40}O_2$				x
	Gadoleic (*cis-9-Eicosenoic*)	$C_{20}H_{38}O_2$				
	Arachidonic (*5, 8, 11, 14-Eicosatetraenoic*)	$C_{20}H_{32}O_2$				
C_{22}	Behenic (*Docosanoic*)	$C_{22}H_{44}O_2$				
	Erucic (*cis-13-Docosenoic*)	$C_{22}H_{42}O_2$				
	Clupanodonic (*4 (?), 8, 12, 15, 19-Docosapentaenoic*)	$C_{22}H_{34}O_2$				
C_{24}	Lignoceric (*Tetracosanoic*)	$C_{24}H_{48}O_2$				
	Nisinic (*4 (?), 8, 12, 15, 18, 21-Tetracosahexaenoic*)	$C_{24}H_{38}O_2$				
C_{26}	Cerotic (*Hexacosanoic*)	$C_{26}H_{52}O_2$				
C_{28} and over	Montanic (*Octacosanoic*)	$C_nH_{2n}O_2$				

X = Trace
Source: Ref. 9.

Cottonseed	Linseed	Oiticica	Olive	Ouri-Curi	Palm	Palm Kernel	Peanut	Perilla	Safflower	Sesame	Soybean	Sugarcane	Sunflower	Tall Oil	Teaseed	Tung
				10		3										
x				9		4						x				
x				46		51						4				
1			x	9	1	17	x		x		x	1				
x									x							
29	6	7	14	8	48	8	6	7	8	10	11	25	11	5	8	4
2	x		2				x		x			1				
4	4	5	2	2	4	2	5	2	3	5	4	4	6	3	1	1
24	22	6	64	13	38	13	61	13	13	40	25	14	29	46	83	8
40	16		16	3	9	2	22	14	75	43	51	36	52	41	7	4
	52		2				x	64	1	2	9	13	2	3		3
																80
		78														
x	x		x				2		x		x	1		2	1	
	x								x		x	1				
							3									
							1									

4%
Hydroxy
acids

Fatty acids 45%
Rosin acids 42%
Terpenes 13%

REFERENCES

1. J. J. Berzelius, *Rapp. Ann.*, *26*:260 (1847).

2. W. H. Carothers, *J. Amer. Chem. Soc.*, *51*:2548, 2560 (1929).

3. R. W. Lenz, *Organic Chemistry of Synthetic High Polymers*, Interscience, New York, 1967, p. 6.

4. P. J. Flory, *J. Amer. Chem. Soc.*, *64*:2205 (1942).

5. P. J. Flory, *Principles of Polymer Chemistry*, Cornell University Press, New York, 1953.

6. H.-G. Elias, *Macromolecules 2, Synthesis and Materials*, Plenum Press, New York, 1977, pp. 602, 615.

7. W. C. Spitzer, *Off. Dig.*, *36*:16 (1964).

8. W. H. Carothers, *Trans. Faraday Soc.*, *32*:39 (1936).

9. D. H. Solomon, *The Chemistry of Organic Film Formers*, Krieger Publ., New York, 1977, pp. 65, 82, 88, 108.

10. C. W. Johnston, *Off. Dig.*, *32*:1327 (1960).

11. L. A. Tysall, *Calculation Techniques in the Formation of Alkyds and Related Resins*, Paint Research Association, Teddington, United Kingdom, 1982, p. 18.

12. A. Gal and A. Hardof, *Org. Coat. Sci. Technol.*, *8*:55 (1986).

13. P. J. Flory, *J. Amer. Chem. Soc.*, *58*:1877 (1936).

14. G. Christensen, *Off. Dig.*, *36*:28 (1964).

15. D. Firestone, *J. Amer. Oil Chem. Soc.*, *40*:247 (1963).

16. D. F. Rushman and E. M. G. Simpson, *Trans. Faraday Soc.*, *51*:230 (1955).

17. P. J. Flory, *Chem. Rev.*, *39*:137 (1946).

18. S. D. Hamann, D. H. Solomon, and J. D. Swift, *J. Macromol. Sci. Chem.*, *A2*:153 (1968).

19. J. H. Saunders and F. Dobinson, in *Comprehensive Chemical Kinetics, Vol. 15: Non-Radical Polymerization* (C. H. Bamford and C. F. H. Tipper, eds.), Elsevier, Amsterdam, 1976, pp. 505, 507.

20. C. E. H. Bawn and M. B. Huglin, *Polymer*, *3*:257 (1962).

21. R. Brown, H. Ashjian, and W. Levine, *Off. Dig.*, *33*:539 (1961).

22. R. A. Brett, *J. Oil Col. Chem. Assoc.*, *41*:428 (1958).

23. R. O. Feuge and A. E. Bailey, *Oil Soap*, *23*:259 (1946).

24. E. F. Pratt and J. D. Draper, *J. Amer. Chem. Soc.*, *71*:2846 (1949).

25. L. M. R. Crawford and D. A. Sutton, *Chem. Ind.*, *38*:1232 (1970).

26. J. T. Geoghegan and W. E. Bambrick, *J. Paint Technol.*, *42*: 490 (1970).

27. G. Walz, *J. Oil Col. Chem. Assoc.*, *60*:11 (1977).

28. H. G. Ramjit, *J. Macromol. Sci. Chem.*, *A19*:41 (1983).

29. V. N. Kursanov, V. V. Korshak, and S. V. Vinogradova, *Bull. Acad. Sci., USSR, Div. Chem. Sci.*, 125 (1953).

30. R. G. Robins, *Australian J. Appl. Sci.*, *5*:187 (1954).

31. R. Hauschild and J. Petit, *Bull. Soc. Chim. France*, 878 (1956).

32. A. Tremain, *J. Oil Col. Chem. Assoc.*, *40*:737 (1957).

33. W. G. Bickford, P. Krauczunas, and D. H. Wheeler, *Oil Soap*, *19*:23 (1942).

34. E. Eslami, *J. Rech. C.N.R.S.*, *61*:333 (1962).

35. J. Ross, A. I. Gebhart, and J. F. Gerecht, *J. Amer. Chem. Soc.*, *68*:1373 (1946).

36. H. M. Teeter, M. J. Geerts, and J. C. Cowan, *J. Amer. Chem. Soc.*, *25*:158 (1948).

37. K. Holmberg and J.-A. Johansson, *Acta Chem. Scand.*, *B36*: 481 (1982).

38. A. E. Rheineck and T. H. Khoe, *Fette Seifen Anstrichm.*, *71*:644 (1969).

39. H. J. Lanson, in *Kirk-Othmer, Encyclopedia of Chemical Technology*, 3rd Ed., Vol. 2, Wiley, New York, 1982, p. 33.

40. E. G. Bobalek, E. R. Moore, S. S. Levy, and C. C. Lee, *J. Appl. Polym. Sci., 8*:625 (1964).

41. R. G. Mraz, R. P. Silver, and W. D. Coder, *Off. Dig., 29*: 256 (1957).

42. J. R. Fletcher, L. Polgar, and D. H. Solomon, *J. Appl. Polym. Sci., 8*:659 (1964).

43. D. H. Solomon and J. J. Hopwood, *J. Appl. Polym. Sci., 10*:993 (1966).

44. H. A. Goldsmith, *Ind. Eng. Chem., 40*:1205 (1948).

45. K. Holmberg and J.-A. Johansson, *Org. Coat. Sci. Technol., 6*:23 (1984).

46. J. P. Walsh, *Mod. Paint Coat., 73*:42 (Nov. 1983).

47. W. M. Kraft, G. T. Roberts, E. G. Janusz, and J. Weisfeld, *Amer. Paint J., 41*:96 (1957).

48. J.-A. Johansson and K. Holmberg, unpublished work.

49. C. B. Reese, in *Protecting Groups in Organic Chemistry* (J. F. W. McOmie, ed.), Plenum Press, New York, 1973.

50. J. Dörffel, *Farbe Lack, 88*:6 (1982).

51. D. H. Solomon and J. J. Hopwood, *J. Appl. Polym. Sci., 10*: 1893 (1966).

52. H. Hata, T. Tomita, and J. Kumanotani, *Proc. FATIPEC Congr.*, Amsterdam, 1980, p. 356.

53. J. Kumanotani, H. Hata, and H. Masuda, *Org. Coat. Sci. Technol., 6*:35 (1984).

54. A. Kastanek, S. Podzimek, and K. Hajek, *J. Appl. Polym. Sci., 30*:4723 (1985).

55. F. M. Ball, in *Treatise on Coatings* (R. R. Myers and J. S. Long, eds.), Vol. 1, Part III, Marcel Dekker, New York, 1972, pp. 296-302.

56. A. E. Rheineck and R. O. Austin, in *Treatise on Coatings* (R. R. Myers and J. S. Long, eds.), Vol. 1, Part II, Marcel Dekker, New York, 1968, p. 199.

4
The Solvent

Selection of solvents for conventional coatings is a relatively straightforward procedure. Since these coatings contain a high level of volatile material, the formulater is allowed a wide latitude in the choice of solvent. As a result, a number of solvents are often employed interchangeably in the formulations, depending on costs.

In high solids systems the amount of solvent allowed is small and the selection of proper solvent combinations is, therefore, crucial. The purpose of the solvent is to control film flow during and shortly after application. With too much or too little flow a poor film develops. One of the differences between high solids and conventional resins is that the former are inherently more fluid. Also, as solvent evaporates, the viscosity of the conventional system increases more rapidly than the high solids system and is higher at each concentration. As a consequence, high

solids coatings will be expected to flow more at all stages of drying. Avoiding excessive flow is, thus, a primary problem with high
solids coatings.

   In this chapter the solvent parameter concept is reviewed
and the mechanism of solvent evaporation from conventional and
high solids systems discussed. The effect of solvent on two coating parameters—viscosity and surface tension—is described.

   The environmental and legislative aspects, although of
major importance in solvent selection, are not specifically dealt
with here since the restrictions undergo frequent changes and,
furthermore, differ from country to country. Toxicology, fire
hazards and related topics are also not included.

4.1   SOLUBILITY PARAMETERS

Hildebrand has defined the solubility parameter, δ, or rather its
square, the cohesive energy density (ced), as

$$\delta^2 = ced = \frac{\Delta E_v}{V}$$

where ΔE_v is the energy of vaporization and V is the molar volume [1]. The basis of the Hildebrand theory resides in the fact
that when the solubility parameters of two materials are equal,
the materials are infinitely soluble. To the extent that solubility
parameters differ, solubility decreases.

   Hildebrand's original theory dealt only with nonpolar
materials, i.e., those in which only dispersion forces act between
the molecules. The theory has been extended, by the efforts of
Hansen and others, to include molecules interacting by dipolar
and hydrogen-binding forces [2-4]. It was assumed that dispersion, polar, and hydrogen-bonding parameters are simultaneously

valid, their particular values being determined by a large number
of experimental solubility observations.

A three-component solubility parameter concept was introduced by making the assumption that the ced can be repre-
sented by an additive function:

$$\delta^2 = \delta_d{}^2 + \delta_p{}^2 + \delta_h{}^2$$

where the subscripts refer to the dispersion, polar, and hydrogen-
bonding contributions, respectively. It has been pointed out that
"hydrogen bonding" may not be adequate in all cases [5]. Some
more ambiguous term, such as "weak chemical bonds" or "association bonds" might be better. However, "hydrogen bonding" is
normally used from historical grounds.

A series of solubility parameter values for solvents of relevance to high solids coatings are given in Table 4.1. The methods
for calculating those components of the ced are given by Hansen
and Beerbower [5], and the conditions for mutual solubility now
require a match for all three components. The original unit of
the solubility parameter, $cal^{1/2}/cm^{3/2}$, is usually designated as a
"hildebrand." The corresponding SI unit is $MPa^{1/2}$, which is equal
to 0.488 hildebrand.

There exist various ways of making graphic representations
of the solubility parameter data, as described in a review [6].
When Hansen's three-dimensional treatment is used, an ellipsoid
is obtained. Expansion of the δ_d scale by a factor of two results
in a convenient spherical solubility volume for a solute. In this
way solubility spheres for individual polymers can be plotted on a
three-dimensional system of coordinates.

In constructing a solubility parameter plot the polymer is
added to a representative selection of solvents to see whether it
dissolves, swells, or remains unaffected. The concentration of the

Table 4.1 Solubility Parameters of Selected Solvents

Solvent	Molar volume (cm^3)	Parameters ($MPa^{1/2}$)		
		δ_d	δ_p	δ_h
n-Dodecane	229	16.0	0	0
Toluene	107	18.0	1.4	2.0
o-Xylene	121	17.8	1.0	3.1
Ethyl methyl ketone	90	16.0	9.0	5.1
Cyclohexanone	104	17.8	6.4	5.1
Methyl isobutyl ketone	126	15.3	6.1	4.1
n-Butyl acetate	133	15.8	3.7	6.3
2-Ethoxyethyl acetate	136	16.0	4.7	10.7
Methanol	41	15.1	12.3	22.3
Ethanol	59	15.8	8.8	19.4
1-Propanol	75	16.0	6.8	17.4
2-Propanol	77	15.8	6.1	16.4
1-Butanol	92	16.0	5.7	15.8
2-Butanol	92	15.8	5.7	14.5
Ethylene glycol monoethyl ether	98	16.2	9.2	14.3
Diethylene glycol monoethyl ether	131	16.2	9.2	12.3
Ethylene glycol mono-n-butyl ether	132	16.0	5.1	12.3
Ethylene glycol	56	17.0	11.1	26.0
Propylene glycol	74	16.8	9.4	23.4
Diethylene glycol	95	16.2	14.8	20.5
Water	18	15.6	16.0	42.4

Source: Ref. 4.

polymer solution is not critical, however concentrations around 10% have been found to be convenient [7]. The solubility range on each axis in the δ_d, δ_p, δ_h system is identified and a sphere can be constructed that encompasses all the good solvents. This is illustrated for polymethylmethacrylate in Figure 4.1.

In many cases a simple δ_p vs. δ_h plot is sufficiently accurate. This simplifies the plotting of the data since a two-dimensional system is obtained. It must be remembered, however, that this approach will give somewhat erroneous results with some solvents which, like most polymers, have high δ_d components. Halogenated and many cyclic solvents fall into this category.

The use of a hydrogen bond parameter, δ_h, to describe the influence of hydrogen bonds on the dissolving capacity has been questioned, however. It has also been claimed that Hansen's method is not reliable since it neglects entropy changes [8].

An alternative way of determining solubility parameters for polymers is to arrange solvents quantitatively into three classes:

poorly hydrogen bonded (e.g., hydrocarbons and halogenated hydrocarbons)
moderately hydrogen bonded (e.g., esters, ethers, glycol monethers, and ketones)
strongly hydrogen bonded (e.g., alcohols, amines, and acids)

Solubility parameters for a given polymer are then determined experimentally for each hydrogen bonded class of solvents. The solvents to be used are conveniently selected from the solvent spectra given in Table 4.2. Further details of the experimental procedure are given in Ref. 9.

Solubility parameter values for a list of commercial polymers are given in the literature [9]. A number of alkyd resins

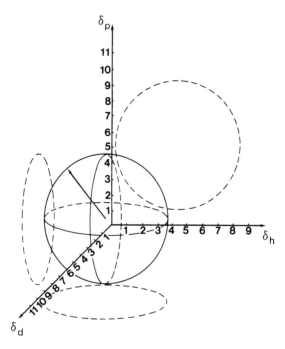

Figure 4.1 Solubility plot for polymethylmethacrylate. The areas within the circles represent a solubility of 10% or more. (From Ref. 5.)

with varying oil length, etc. are also included. Using swelling measurements the solubility parameters of a series of unsaturated polyester resins were determined [10]. It was found that the solubility parameters decreased with increasing degree of cross-links.

Recently, a new method for measuring solubility parameters of polymers has been proposed. Two polymer samples of the same composition but with different molecular weights are required. Gel permeation chromatography and intrinsic viscosity determinations provide the data needed. The polymer solubility parameter is conveniently obtained from the slope of a plot [11].

Table 4.2 Solvent Spectra

Solvent	δ (MPa$^{1/2}$)
Poorly Hydrogen Bonded	
n-Pentane	14.3
n-Heptane	15.1
Methylcyclohexane	16.0
Solvesso 150	17.4
Toluene	18.2
Tetrahydronaphthalene	19.4
o-Dichlorobenzene	20.5
1-Bromonaphthalene	21.7
Nitroethane	22.7
Acetonitrile	24.2
Nitromethane	26.0
Moderately Hydrogen Bonded	
Diethyl ether	15.1
Diisobutyl ketone	16.0
N-Butyl acetate	17.4
Methyl propionate	18.2
Dibutyl phthalate	19.1
Dioxane	20.3
Dimethyl phthalate	21.9
2,3-Butylene carbonate	24.8
Propylene carbonate	27.3
Ethylene carbonate	30.1
Strongly Hydrogen Bonded	
2-Ethyl hexanol	19.5
Methyl isobutyl carbinol	20.5
2-Ethylbutanol	21.5
n-Pentanol	22.3
n-Butanol	23.4
n-Propanol	24.4
Ethanol	26.0
Methanol	29.7

Source: **Ref. 8.**

Determination of solubility parameters for a polymer has proved to be extremely useful in selecting the proper solvent combination. For instance, a mixed, good solvent, composed exclusively of nonsolvents, can be obtained by choosing nonsolvents located, respectively, on opposite sides of the region of solubility. Questions relating to solvency, polymer compatibility, and swelling of polymers can easily be visualized by this approach. Thixotropy of alkyds, which is usually caused by the incorporation of small amounts of hydrocarbon-insoluble material into the polymer at a late stage of the synthesis, may also be approached by the solubility parameter concept [5].

Thus, the solubility parameter concept is a tool to determine possible combinations of solvents that will solve a particular solubility problem. This type of systematic approach to solvent selection is particularly important in formulating high solids coatings.

An alternate way of quantifying the polar interactions between a solvent and a solute has been proposed by Fowkes [12,13]. Lewis' acid-base concept is the basis of the approach; hence, the acid is an electron acceptor and the base is an electron donor. The strength of the acid and the base is governed by

1. the ability to accept or donate an electron pair
2. the degree of polarizability

The enthalpy of interaction between an acid and a base can be written [14]:

$$-\Delta H_{AB} = C_A C_B + E_A E_B$$

Where C_A and E_A are empirically determined parameters, assigned to each acid, and C_B and E_B are assigned to each base such that when substituted into the above equation, they give the enthalpy of adduct formation for the acid-base pair. An extensive set of C and E parameters has been collected [14].

For the majority of acids and bases the above equation gives values for $-\Delta H_{AB}$ which are within ± 0.4 kJ/mol from the experimentally determined ones. In this treatment hydrogen bondings are seen as just one example of acid-base interactions. Hence, according to Fowkes' theory, only two modes of interaction between molecules produce the cohesive energy characteristic of the liquid state—nonpolar dispersion forces and acid-base interactions.

The concept of acid-base interactions seems to be particularly useful in illustrating the competing interaction between solute and solvent and between solute and a solid surface (adsorption), respectively. An example is given in Figure 4.2, showing the amount of polymethylmethacrylate (PMMA) adsorbed on silica from different solvents as a function of acidity or basicity of the solvent [13]. The latter parameter is characterized by ΔH_{AB} for the solvent calculated against ethyl acetate and butanol, respectively. Strong adsorption is obtained from neutral solvents (benzene, tetrachloromethane, dichloromethane) since PMMA is a basic polymer and silica is an acidic surface. Dioxane and tetrahydrofuran are strongly basic solvents and are, therefore, associated so strongly to the surface that the polymer is unable to displace them. Chloroform, on the other hand, is strongly acidic and is so strongly bonded to the polymer in solution that the tendency for polymer adsorption is small. On basic surfaces, such as calcium carbonate, there is very little adsorption of PMMA from any of the solvents.

4.2   SOLVENT EVAPORATION

Neat solvent evaporation is a straightforward phenomenon. It depends upon vapor pressure, which, in turn, varies with temperature. For solvent blends vapor pressure also varies with molecular interactions.

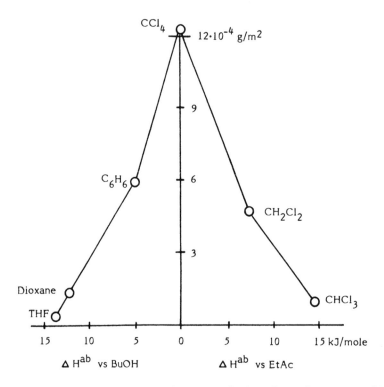

Figure 4.2 Adsorption of polymethylmethacrylate onto silica from basic solvents (left) and from acidic solvents (right). (From Ref. 13.)

The situation becomes much more complex when polymer solutes are involved. From such solutions the solvent evaporation rate is considerably reduced, especially at the later stages when the solute concentration is high. An illustration of this behavior is shown in Figure 4.3. The difference in evaporation time for a solvent with and without a polymer solute is more pronounced for the faster evaporating solvents. Whereas heavy aromatics and paraffinic hydrocarbons show a time ratio solution: neat solvent for 90% evaporation of 2-3, more volatile solvents may give a ratio of more than 10 [15].

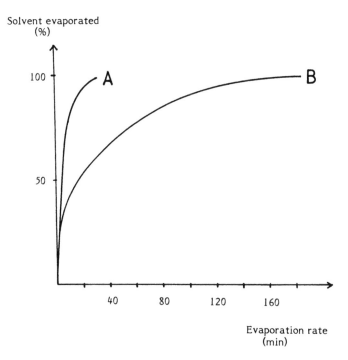

Figure 4.3 Solvent evaporation plots of (A) neat solvent and (B) 40% solution of a long oil alkyd resin. Methylcyclohexane was used as solvent. (From Ref. 15.)

It has been shown that two factors are of particular importance with regard to solvent retention: the degree of branching of the solvent molecule and the formation of strong association, notably hydrogen bonds, between the solvent and the solute [15]. A comparison of the 90% evaporation time ratios for *n*-butanol and isobutanol nicely illustrates the effect of branching. The straight chain isomer has a time ratio of 5, i.e., it takes 5 times longer for 90% of the solvent to evaporate from a resin solution than for the neat solvent, and the branched isomer gives a ratio of 9. Generally, the larger the cross-sectional area of the solvent, the more pronounced is the retardation of evaporation.

As is evident from Figure 4.3, evaporation of solvent from a polymer solution is a two-stage process. According to Raoult's law, in an ideal system the partial pressure of the solvent is reduced proportionally to the mole fraction of the polymer solute. However, due to its high molecular weight a resin exerts little effect. Therefore, the partial pressure of the solvent in the early stages of drying is similar to the vapor pressure of the neat solvent. This holds true for the system long oil alkyd-methylcyclohexane shown in Figure 4.3.

The second evaporation stage of a solvent from a polymer solution is not simply a function of solvent volatility. Instead, diffusion of the solvent molecules to the surface where they can evaporate is the rate-limiting process [16]. The diffusion process is dependent upon size and shape of the solvent molecule, and the relative rate of diffusion of a series of solvents will remain the same regardless of the type of polymer [17]. The diffusion coefficient of a single solvent is, of course, dependent on the type of polymer used. For instance, diffusion is more rapid with a resin when it is above its glass transition temperature than when it is below it. Plasticizing a resin also increases the rate of solvent diffusion.

It is a well-known fact that many resins, especially the thermoplastic ones, may retain considerable amounts of solvent in the film for long periods of time. Retention depends upon the choice of both solvent and resin. Hard resins retain solvent to a higher degree than more fluid ones, and plasticizing of the resin reduces solvent retention [17,18].

The point of transition from volatility-controlled to diffusion-controlled evaporation is of considerable importance to the paint formulater since this represents the stage where the rate of solvent loss slows down significantly. The term "specific evaporation rate" has been introduced as a measure of the retarding power of the resin [15]. For a given percent solvent evaporated, the specific evaporation rate is the ratio of the evaporation rate decrease caused by the resin to the evaporation rate for the neat solvent:

$$\text{specific evaporation rate} = \frac{\dfrac{dw}{dt}(\text{soln}) - \dfrac{dw}{dt}(\text{solv})}{\dfrac{dw}{dt}(\text{solv})}$$

where

w = weight percent solvent evaporated

t = time

A plot of the specific evaporation rate as a function of volume percent resin for the long oil alkyd in methylcyclohexane mentioned earlier is shown in Figure 4.4. The transition from volatility-controlled to diffusion-controlled evaporation is very apparent. The value chosen as the transition point is the volume percent resin where the specific evaporation rate equals —0.50.

The transition point is influenced both by solvent and resin types. Generally, the transition point occurs at higher resin concentrations as solvent volatility decreases. Going from n-heptane to n-decane, for instance, raises the transition point of a specific alkyd solution from 38 to 55 volume percent [15].

Comparisons between isomeric linear and branched solvent molecules show that whereas the branched ones usually have a higher rate of evaporation before the transition point, they are more retarded when diffusion is the rate-limiting factor. Solvents with strong hydrogen bonding capability also give low rate of evaporation after the transition point.

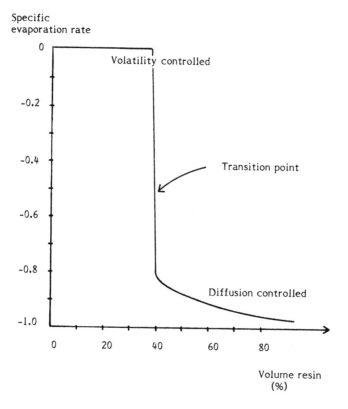

Figure 4.4   Specific evaporation rate vs. volume % resin for the alkyd solution of Figure 4.3. (From Ref. 15.)

Evaporation measurements with a series of saturated polyesters clearly show that the transition point occurs at higher polymer concentration for high solids resins than for conventional ones [15]. This is illustrated in Figure 4.5. Resin A, which has the lowest molecular weight and the most narrow molecular weight distribution, has the highest transition point.

In the formulation of a paint it is extremely important whether or not the resin percentage is below the transition point. The setting of a coating occurs as viscosity increases rapidly due

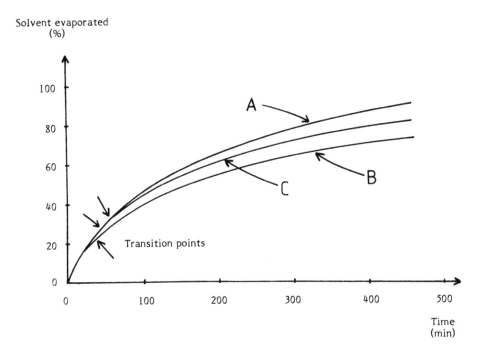

Figure 4.5 Evaporation of 2-ethoxyethyl acetate (ethyl glycol acetate) from three polyester resins. Acid value, Mn, Mw/Mn, and viscosity (in Gard.-Holdt) for resins A, B, and C are: 15, 785, 1.7, and 20-22; 10, 1004, 1.9, and 24-26; 6, 1974, 6.7, and 24-26. (From Ref. 15.)

to fast solvent evaporation. This rapid increase in solids content of the film can only take place if solvent volatility, not diffusion, is the rate-limiting factor. This means that if the formulation is above the transition point, there is a clear risk that solvent evaporation will be too slow to control setting. Excessive flow leading to sagging and nonuniformity in the film may develop.

Conventional alkyd resins are usually formulated so that the vehicle resin percentages are well below the transition point. With high solids alkyd resins this may create a problem, however.

In many cases the resin concentration in the formulation lies above the transition point, implying that solvent evaporation throughout the entire drying/curing process is diffusion controlled. Thus, in many high solids formulations solvent volatility cannot be used to control flow properties. This is a serious problem with these systems and has to be compensated for by some other formulation approach if high solids coatings are going to perform satisfactorily [19-21].

4.3 VISCOSITY

High solids formulations can only be achieved with the proper choice of solvents for the specific polymer in question. To obtain low viscosity or, alternatively, high solids content at a given viscosity, solvents with good solvency must be employed. To this end, the solubility parameter approach is widely used.

High solids resins, due to their relatively low molecular weight, generally have a greater number of solvents capable of dissolving them, i.e., their solubility regions are larger than those of conventional resins. This, however, does not mean that solvent selection for high solids systems is simple. The comparatively small amount of solvent that goes into the formulation is the main tool to achieve such diverse properties as low viscosity and acceptable flow characteristics of the paint.

As mentioned in Section 4.2, a high solids coating is usually formulated above the transition point between volatility-controlled and diffusion-controlled evaporation of the solvent. This means that high solids systems lack the rapid initial increase in solids content after application with a concomitant steep increase in viscosity which is exhibited by conventional systems.

One way of improving viscosity control of high solids coatings is to use combinations of solvents with different solvency and evaporation characteristics. The solubility parameter approach is applicable and the principle is outlined in Figure 4.6. The initial solvent composition is designed to give maximum solvency, i.e., the lowest possible viscosity. Solvent evaporation leads to a gradual deterioration of solvency, shown in Figure 4.6 as a change in composition from the interior to the boundary of the solubility region of the polymer. The transition from a good to a poor solvent composition will, of course, have a major influence on viscosity.

The above approach may be thought of as contrary to that taken in many formulations of conventional systems. In these, aliphatic or aromatic hydrocarbons are normally used extensively, so that the resulting average solubility will be in the lower left-hand portion of the solubility region very close to the boundary. The solvents are chosen so that the poorest solvents evaporate first, thus moving the composition inward, away from the boundary. It is obvious that control of polymer precipitation and pigment flocculation is easier by this approach than by the one described above for high solids systems.

4.3.1 Influence of Resin Concentration

The greatest resin-solvent interaction occurs when the solvent and the polymer have the same solubility parameters. At this point the polymer molecules will have uncoiled to the maximum. At low concentrations this will lead to high viscosities [22]. At higher concentrations the situation is reversed: entanglement of polymer chains governs the solution viscosity and thermodynamically good solvents permit slippage at entanglement points, resulting in lower viscosities [23,24]. Hence, selecting solvents having

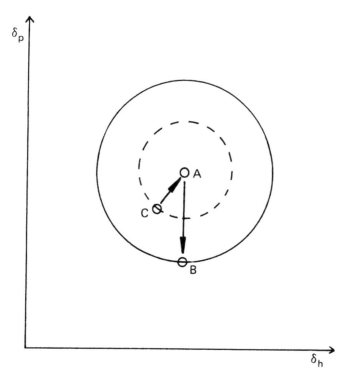

Figure 4.6 Schematic illustration of change in solvent composition during evaporation. The circles show regions of solubility for a high solids resin (full line) and a conventional resin (broken line). For the high solids coating the transition goes from A to B; for the conventional coating it goes from C to A.

the same solubility parameter as the binder (or the binder combination) would appear to be the key to low viscosity formulations.

   With high solids alkyd resins the situation is more complicated than that, however. The resins in many cases have molecular weights below the critical molecular weight of entanglement. Due to too low a degree of polymerization the random coil model may not be an accurate model of the chain conformation. For these resin systems polymer-solvent interactions are extremely complex and seem not to be fully understood.

Erickson studied oligomers of molecular weight 1,000-3,000 at the concentrations used in high solids coatings. He found that for a large number of polymer-solvent systems viscosity vs. concentration followed the relationship [25]:

$$\frac{W}{\ln(\eta/\eta_s)} = \frac{1}{[\eta]_w} + K_w \cdot W \qquad (4.1)$$

where

W = concentration of oligomer
η = solution viscosity
η_s = solvent viscosity
$[\eta]_w$ = weight intrinsic viscosity
K_w = a constant

He determined experimentally that the term $[\eta]_w$ showed a relatively high value at the center of the solubility volume and a minimum roughly halfway between the center and the edge. It was stated that $[\eta]_w$ depends on the difference in solubility parameter between the polymer and the solvent ($\Delta\delta$). When $\Delta\delta$ is zero, the resin molecules attain an extended conformation due to maximum interaction with the solvent. This leads to high $[\eta]_w$-values and, consequently, to high values of η. As $\Delta\delta$ increases, oligomer segments begin to interact intramolecularly, which causes a decrease in $[\eta]_w$. The increase in $[\eta]_w$, which occurs on further $\Delta\delta$ increase, has been attributed to intermolecular interaction of resin molecules with the formation of clusters.

Erickson's original treatment, which involved a single resin in a single solvent, was later successfully extended to two-resin systems [26]. Attempts to extend this semi-empirical concept to more complex solvent systems were not entirely successful, however [27]. The $[\eta]_w$-values obtained with mixtures of a good and a poor solvent at various ratios were all almost the same as that

obtained with the good solvent alone. Furthermore, using a high
solids saturated polyester it has been shown that pure solvents
and solvent mixtures, having the same solubility parameters, can
give very different effects on the solution viscosities, even if influences of densities and inherent viscosities of the solvents are compensated for [28]. Evidently, the good solvent retains its solvency
also in a mixture, thus making $[\eta]_w$ (and η) practically independent of $\Delta\delta$ ($\delta_{polymer} - \delta_{solvent\ mixture}$).

Since high solids alkyd resins are almost invariably formulated with solvent mixtures, Erickson's concept seems to be of
limited practical use. However, it could serve as a starting point
for further research in the field of polymer-solvent interactions in
high solids systems. Evidently, a better understanding of the fundamental phenomena involved in this area is needed.

4.3.2 Solvent Viscosity

The viscosity of a resin solution is strongly dependent on the viscosity of the solvent itself. This is illustrated in Figure 4.7 for
solutions of polystyrene in a series of solvents of varying viscosities. There is a straight-line relationship between the viscosity of
the solvent and that of the polystyrene solution [29].

A general equation to relate solution and solvent viscosities for high solids systems has been proposed [30]:

$$\log\eta = \log\eta_s + \frac{W}{K_a - K_b \cdot W} \tag{4.2}$$

where η, η_s, and W are the same as in Eq. (4.1) and K_a and K_b are
constants.

It is evident from Eq. (4.2) that although the solution viscosity, η, is always dependent on η_s, the relative importance of
this term decreases as the resin concentration, W, increases. For

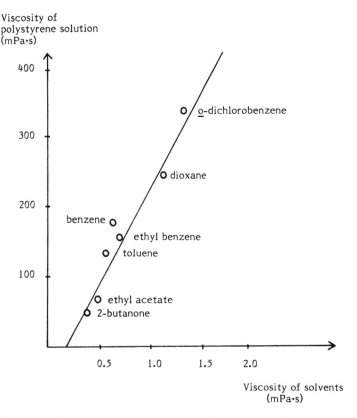

Figure 4.7   Viscosity of a 12% polystyrene solution in solvents of varying viscosities. (From Ref. 29.)

high solids systems the second term of Eq. (4.2) becomes the more important one and factors that control K_a and K_b, such as specific interactions (see Section 4.3.3) and differences in solubility parameters between polymer and solvent (see Section 4.3.1), are decisive [24].

Solvents having low neat viscosities are, consequently, suitable for high solids formulations. By the same token, solvents having high viscosities are usually unsuitable for this purpose. This is the main reason why linseed oil, otherwise the ideal reactive diluent for air-drying alkyds, have not gained widespread use as a solvent (see Section 6.1).

Ketones such as methyl isobutyl keton (4-methyl-2-pentanone), methyl ethyl ketone (2-butanone), and cyclohexanone constitute a class of solvents that combine good solvency and low neat viscosity. Consequently, they are attractive solvents for high solids formulations [31,32]. Chlorinated solvents also have low viscosities and have been suggested for high solids coatings [33]. Although the use of chlorinated hydrocarbons has been put forward as a means of avoiding legislative control, their disadvantages in terms of toxicology and environmental effects are obvious.

4.3.3 Solvents as Hydrogen Bond Acceptors

It has been empirically found that addition of small amounts of solvents that can act as hydrogen bond acceptors in many cases are effective for reducing the viscosity of high solids systems. This effect cannot be explained by the solubility parameter or the solvent viscosity concepts only.

Hydrogen bond acceptors, such as ketones, are more effective in this respect than hydrogen bond donors. Solvents that can act as both acceptors and donors of hydrogen bonds, such as alcohols, are less effective than "acceptor only" solvents. Probably, the latter type of solvent acts by eliminating intermolecular polymer interactions by blocking hydroxyl and carboxyl functional groups [24,34].

4.3.4 Solvent Density

As a result of the way solvents are limited in the United States (weight of solvent by volume of coating), the density of the solvent is an important factor in the formulation of high solids coatings. From this point of view, low density solvents, such as ke-

tones, alcohols, and aliphatic hydrocarbons, are particularly useful.

4.4 SURFACE TENSION

The wetting of the surface by the coating is a fundamental phenomenon for all paints and lacquers. The wetting process may be
visualized as an increase of the solid-liquid and liquid-gas interfacial areas and a decrease of the solid-gas interfacial area. The
well-known expression

$$\gamma_{SL} + \gamma_{LG} \cdot \cos\theta - \gamma_{SG} = 0$$

which is referred to as Young's equation, shows that, at equilibrium, the contact angle, θ, is dependent upon all three interfacial
tensions (see Fig. 4.8).

It can be shown that, for nonpolar liquids, Young's equation can be transformed into the following expression:

$$\cos\theta = -1 + 2\left(\frac{\gamma_s^d}{\gamma_{LG}}\right)^{1/2}$$

where γ_s^d is the dispersion force contribution of the surface tension
of the solid surface, and γ_{LG} is the liquid-gas interfacial tension.
For a series of nonpolar liquids it follows that θ should decrease as
γ_{LG} decreases and become zero below a certain value of γ_{LG}.
Zisman has named this value of γ_{LG} the critical surface tension, γ_c,
for the solid [35]. Critical surface tension is a useful parameter
for characterizing the wettability of solid surfaces.

Surface tension manifests itself in many ways in coatings.
In order for a coating to spread on a substrate, the surface tension
of the liquid coating must be lower than the critical surface tension of the substrate. In addition, since a liquid is easier to break
up and to atomize when the surface tension is low, lower surface

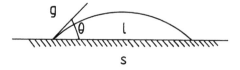

Figure 4.8 A drop of liquid (l) in contact with a solid surface (s) and a gas (g). θ = contact angle.

tension means better sprayability of the coating. The well-known coatings defect, "cratering," is also surface tension related. Contaminants on the surface with very low surface tension, such as fingerprints and oil spots, dissolve in the coating, thus creating areas with lower surface tension than the surrounding coating. The result is that the coating in the low surface tension spots is pulled away into the surrounding higher surface tension area, creating a "crater." The phenomenon, well-known in surface chemistry, is an example of the Marangoni effect.

The surface tension of the coating is largely determined by the polymer and the solvent. The polymers usually have relatively high surface tension, values between 35 and 45 mN/m being typical. The solvents have surface tension values between 20 and 30 mN/m (see Table 4.3).

In general, when comparing solvents within the same solvent class it has been found that faster evaporating solvents usually have lower surface tension values than the slower evaporating counterparts. Furthermore, increased branching of the solvent molecule leads to a lowering of the surface tension [31,32]. Isopropyl alcohol, which fulfils the two above-mentioned demands, has an extremely low surface tension, 21.4 mN/m. Evidently, the higher the solvent content of the coating, the lower the surface tension (unless the solvent is water) [36]. Conventional coatings normally lie in the range of 25-32 mN/m. This is low enough

Table 4.3 Surface Tension of Selected Solvents at 20°C

Solvent Type	Surface Tension (mN/m)
Alcohols	21.4-35.1
Esters	21.2-28.5
Ketones	22.5-26.6
Glycol ethers	26.6-34.8
Glycol ether esters	28.2-31.7
Aliphatic hydrocarbons	18-28
Aromatic hydrocarbons	28-30
Water	72.7

Source: Ref. 23.

to give proper wetting on most surfaces, i.e., to get below the values of the critical surface tension of the substrates (see Table 4.4).

As the resin becomes the major component of the coating, problems related to surface tension are frequently encountered. High solids coatings are, therefore, extremely sensitive to dirt and fingerprints (the surface tension of which is around 24 mN/m) [36,37]. Paint defects, such as "cratering" and "picture framing," occurs more frequently with high solids than with conventional systems.

Table 4.4 Critical Surface Tension of Substrates (20°C)

Solid Surface	γ_c (mN/m)
Polytetrafluoroethylene	18
Polyethylene	31
Polystyrene	33
Poly(hexamethylene adipamide)	46

Source: Ref. 38.

To summarize, the relatively high surface tension of high
solids coatings

requires particularly well-prepared substrates
may influence sprayability in a negative way
can give film defects

REFERENCES

1. J. H. Hildebrand and R. L. Scott, *Regular Solutions*, Prentice-Hall, Englewood Cliffs, N.J., 1962.

2. C. M. Hansen, *J. Paint Technol.*, *39*:104,505 (1967).

3. C. M. Hansen and K. Skaarup, *J. Paint Technol.*, *39*:511 (1967).

4. J. D. Crowley, G. S. Teague, and J. W. Lowe, *J. Paint Technol.*, *38*:269 (1966) and *39*:19 (1967).

5. C. M. Hansen and A. Beerbower, in *Kirk-Othmer, Encyclopedia of Chemical Technology*, 2nd Ed., Suppl. Vol., Wiley, New York, 1971, p. 889.

6. A. F. M. Barton, *Chem. Rev.*, *75*:731 (1975).

7. C. M. Hansen, *Chem. Technol.*, *2*:547 (1972).

8. P. L. Huyskens and M. C. Haulait-Pirson, *J. Coat. Technol.*, *57*: 57 (1985).

9. H. Burrell, in *Polymer Handbook*, 2nd Ed., Interscience, New York, 1975, p. IV-337.

10. S. Takahashi, *J. Appl. Polym. Sci.*, *28*:2847 (1983).

11. C. M. Kok and A. Rudin, *J. Coat. Technol.*, *55*:57 (1983).

12. F. M. Fowkes and M. A. Mostafa, *Ind. Eng. Chem. Prod. Res. Dev.*, *17*:3 (1978).

13. F. M. Fowkes, in *Microscopic Aspects of Adhesion and Lubrication* (J. M. Georges, ed.), Elsevier, New York, 1982, p. 119.

14. R. S. Drago, G. C. Vogel, and T. E. Needham, *J. Amer. Chem. Soc.*, *93*:6014 (1971).

15. W. H. Ellis, *J. Coat. Technol.*, *55*:63 (1983).

16. C. M. Hansen, *Off. Dig.*, *37*:57 (1965).

17. D. J. Newman and C. J. Nunn, *Prog. Org. Coat.*, *3*:221 (1975).

18. J. E. Weigel and E. G. Sabino, *J. Paint Technol.*, *41*:81 (1969).

19. D. R. Bauer and L. M. Briggs, *J. Coat. Technol.*, *56*:87 (1984).

20. D. G. Miller, W. F. Moll, and V. W. Taylor, *Mod. Paint Coat.*, *73:53* (April 1983).

21. R. E. Quinn and J. S. Perz, *Amer. Paint Coat. J.*, *69*:36 (April 15, 1985).

22. Y. Isono and M. Nagasawa, *Macromol.*, *13*:862 (1980).

23. K. S. Gandhi and M. C. Williams, *J. Appl. Polym. Sci.*, *16*: 2721 (1972).

24. L. W. Hill and Z. W. Wicks, Jr., *Prog. Org. Coat.*, *10*:55 (1982).

25. J. R. Erickson, *J. Coat. Technol.*, *48*:58 (1976).

26. A. W. Garner and J. R. Erickson, *ACS Org. Coat. Plast. Chem. Prepr.*, *39*:401 (1978).

27. J. R. Erickson and A. W. Garner, *ACS Org. Coat. Plast. Chem. Prepr.*, *37*:447 (1977).

28. D. Stoye and J. Dörffel, *Pigm. Resin Technol.*, *9*:4 (July 1980).

29. H. Burrell, *Off. Dig.*, *29*:1159 (1957).

30. T. C. Patton, *Paint Flow and Pigment Dispersion*, 2nd Ed., Wiley-Interscience, New York, 1979, Chap. 4.

31. G. Sprinkle, *J. Coat. Technol.*, *53*:67 (1981).

32. G. Sprinkle, *Mod. Paint Coat.*, *73*:44 (April 1983).

33. V. L. Stevens and R. H. Lalk, *Mod. Paint Coat.*, *68*:59 (Sept. 1978).

34. J. M. Butler, R. E. Wolf, C. J. Ray, and G. L. McKay, *ACS Symp. Ser.*, *132*:116 (1980).

35. W. A. Zisman, *Adv. Chem. Ser.*, *43*:1 (1964).

36. C. M. Hansen, *Prog. Org. Coat.*, *10*:331 (1982).

37. Z. W. Wicks, Jr., *Amer. Paint Coat. J.*, *68*:30 (April 30, 1984).

38. N. L. Jarvis and W. A. Zisman, in *Kirk-Othmer, Encyclopedia of Chemical Technology*, 2nd ed., Vol. 9, Wiley, New York, 1966, p. 707.

5

Cross-Linking With Amino Resins

Curing of alkyd resins with amino resin cross-linkers leads to the formation of a three-dimensional network. The cross-link density of the network governs film properties, such as hardness, flexibility, adhesion, and durability.

The cross-linking agent influences the film performance of high solids coatings to a greater extent than conventional coatings. This is due to the high solids alkyds often having lower molecular weights than normal alkyds, thus requiring larger levels of cross-linker for the curing process.

The difference in cross-link density between a high solids system and a conventional one is schematically illustrated in Figure 5.1. Normally, the molecular weights of both the alkyd and the amino resin are considerably reduced for high solids systems. If, in the extreme case, the molecular weights of both components are reduced to one-third of their original values, the theoretical

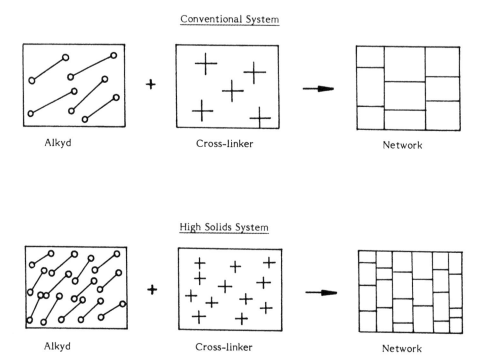

Figure 5.1 A schematic representation of the difference in network build-up between a conventional and a high solids alkyd-cross-linker system. (From Ref. 1.)

number of cross-links per unit volume (three-dimensional network) increases by a factor of more than 30 [1]. Hence, the number of chemical reactions taking place during the curing of high solids systems is much higher than for conventional systems. On the other hand, the number of reaction steps involved in the syntheses of high solids resins is smaller than for normal resins. Thus, the chemistry of the curing process has grown in importance with the introduction of high solids systems. A good knowledge of the structure of the amino resin cross-linker, as well as of the high solids alkyd resin, is, therefore, important. An understanding of the cocondensation and self-condensation reactions involved in the formation of the three-dimensional polymer network is also vital.

In principle, there exists a choice of amino resin type for use in combination with the high solids alkyd resin. In practice, however, melamine resins are almost invariably used for this purpose, and the following presentation is restricted to this type of cross-linking agent. The chapter starts with a short presentation of melamine resins, followed by an outline of the chemistry involved in their curing with alkyds in general. The last section is devoted to the special features encountered with high solids alkyd-melamine resin curing.

5.1 MELAMINE RESINS

Melamine resins are condensation products of melamine, formaldehyde, and a low molecular weight alcohol. They are sometimes referred to as melamine-formaldehyde resins or triazine resins (the IUPAC name for melamine is 2,4,6-triamino-1,3,5-triazine).

$$NH_2$$

2,4,6-Triamino—1,3,5-triazine or melamine

Melamine resins bearing anionic charges, e.g., from sulphonate substituents, or permanent cationic charges exist, but such products are not used as paint binders and are beyond the scope of this monograph. The melamine resins of interest to the paint and lacquer field are those having hydroxyl and primary, secondary, and tertiary amino groups as reactive sites.

In the preparation of melamine resins, melamine is first reacted with formaldehyde to form methylolmelamine. Depending on the ratio between formaldehyde and melamine, condensation products having from one to six hydroxymethyl groups per triazine ring may be obtained. The methylolated melamine compound is usually not isolated, but reacted further with an alcohol, usually either methanol or butanol, to give a fully or partially etherified product. The etherification step is decisive for the solubility characteristics of the resin. Butylated products are usually readily soluble in most organic solvents of interest for paint formulations but insoluble in water. The methylated melamine resins have a broader solubility range and can be used in aqueous, as well as in nonaqueous, systems. The nonetherified melamine-formaldehyde resin is difficult to dissolve both in water and in most organic solvents. The highly methylolated products of this type need polar, aprotic solvents, like dimethyl sulphoxide, to go into solution.

During both the methylolation and the etherification steps a certain degree of condensation, leading to dimers, trimers, etc., takes place. The substituted triazine rings become cross-linked, mainly by methylene diamine bridges ($N-CH_2-N$). The degree of polymerization can to some extent be governed by proper choice of reaction conditions. High reaction temperature, long reaction time, and a low pH during the etherification favor polymerization. A typical melamine-formaldehyde resin for conventional solventborne formulations is in the molecular weight range of 800-2000, having 3-7 condensed triazine units. The reactions involved in melamine resin synthesis are shown in Figure 5.2.

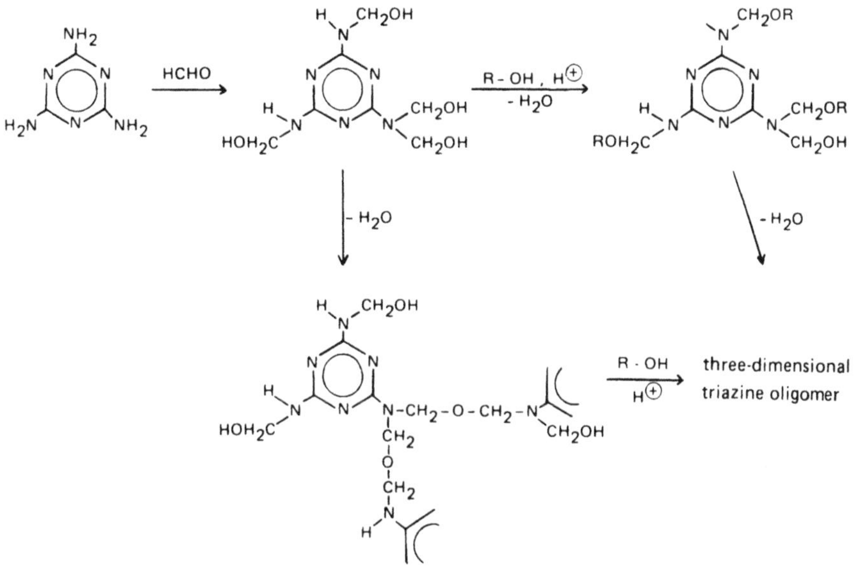

Figure 5.2 Synthesis of a melamine resin.

Melamine resins are normally dissolved in the alcohol used
for etherification. It some cases it is considered advantageous to
use different alcohols for alkylation and as solvent. If the solvent
alcohol has the higher boiling point, this is easily achieved simply
by adding the heavier alcohol to the ready-made resin solution
while distilling off the more volatile one. This procedure will
invariably lead to a product having a mixture of alkoxyl groups
since some transetherification will take place even at neutral pH.

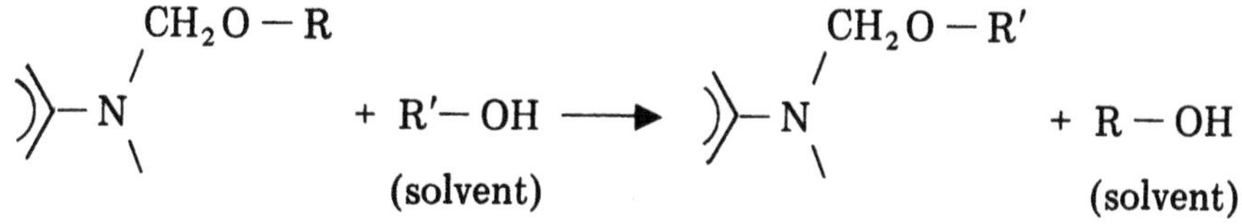

As mentioned above, the most widely used melamine resins are the butylated or methylated ones. Conventional butylated

resins are oligomeric in nature and prepared at approximately 50% solids content. The methylolation is performed at relatively high pH and the subsequent reaction with alcohol is made under acidic conditions.

In order to prepare lower molecular weight butylated melamine resins, processing changes have to be made. The entire reaction is preferably carried out on the acidic side. A one-hour reflux at pH 6 will allow reaction with formaldehyde to take place. The pH is then lowered in order to promote acid-catalyzed butylation.

The equilibrium of the alkylation step is shifted to the right by a continuous azeotropic removal of the water of condensation. The etherification time must be kept as short as possible in order to minimize triazine ring cross-linking. A typical procedure for the preparation of a high solids butylated melamine-formaldehyde resin is given in Table 5.1.

For methylated melamine resins there exists no convenient azeotrope to remove the water formed in the condensation reaction. Hence, a different processing technique than for the butylated resins has to be employed.

In order to minimize the amount of water introduced with the starting materials the solid polymer paraformaldehyde (polyoxymethylene) is normally used as the formaldehyde source. The pH of the mixture must first be raised, preferably by a water-free base, such as a methylate, to dissolve, or rather depolymerize, the paraformaldehyde. The formaldehyde generated readily reacts with the melamine amino groups forming the desired methylol compound. Etherification is then performed by lowering pH. Since the equilibrium of the methylation step cannot be shifted to the right by azeotropic distillation, a large excess of alcohol is needed in order to keep triazine ring cross-linking at a minimum.

Table 5.1 Preparation of a High Solids Butylated Melamine Resin

Components		
Melamine	1 mol	126 parts
Formaldehyde (as 100%)[a]	6 mol	180 parts
n-Butanol	6 mol	444 parts
Formic acid (90%)	as required	

Procedure
1. Charge melamine, formaldehyde, and n-butanol.
2. Adjust pH to 6.0 with formic acid.
3. Reflux for 1 hr.
4. Cool to approximately 80°C. Adjust to pH 4.5 with formic acid.
5. Azeotrope water until desired solids content is reached (about 70%).

[a]Could for instance be added as a formaldehyde solution in n-butanol/water.
Source: Ref. 2.

After completed reaction, excess methanol and formaldehyde are removed and recirculated to the next batch. A typical processing procedure is given in Table 5.2.

A number of low molecular weight methylated melamine resins are available on the market. These are normally referred to as HMMM resins (hexamethoxymethyl melamine), although analysis shows that they are never true monomers. Most HMMM resins seem to have an average molecular weight corresponding to a triazine dimer rather than a monomer.

The melamine resins can be conveniently analyzed with regard to functional groups present by spectroscopic methods. ^{1}H- and ^{13}C-NMR have proved to be excellent techniques both for qualitative and quantitative determinations. Alkoxy groups, as well as most imino and methylol groups, can be analyzed with a high repeatability [3,4].

Table 5.2   Preparation of a High Solids Methylated Melamine
Resin

Components		
Melamine	1 mol	126 parts
Paraformaldehyde (92%)	8 mol	260 parts
Methanol	11 mol	352 parts
Sodium methylate	as required	
Formic acid (90%)	as required	

Procedure
1. Charge melamine, paraformaldehyde, and methanol.
2. Adjust pH to 8.8 with sodium methylate.
3. Heat to 65°C. Hold until all paraformaldehyde dissolves.
4. Adjust pH to 5.3 with formic acid.
5. Continue cooking at 65°C for approximately 7 hr.
6. Cool to room temperature and adjust pH to 7.8.
7. Vacuum strip until desired solids content is reached.

Source: Ref. 2.

The preparation of the melamine resins, as well as their curing with hydroxyl-containing polymers, can be conveniently monitored by IR spectroscopy. For instance, the characteristic adsorption band of the methyl ether appears at 1080 cm^{-1} and that of the methylol group is at 1020 cm^{-1}. The ratio of the two peaks gives a measure of the degree of etherification [2]. The Fourier transform technique (FTIR), with its high signal-to-noise ratio, seems to be particularly suitable to follow these dynamic processes. Solid state ^{13}C-NMR has also been used to study the curing reactions. The data obtained by this technique seem less valuable than those obtained by FTIR, however [5].

Special attention should be paid to the determination of solids content of the methylated amino resins. Temperatures of 100°C or higher, which are normally used for amino resin analyses, may give misleading results due to partial decomposition and, possibly, further condensation of the methylated resin.

Instead, it has been suggested that the nonvolatile content should be measured after 45 min at 45°C [6].

5.2 GENERAL THEORY

The chemistry of the curing of melamine resin systems has been the subject of a vast amount of literature [7,8]. The first detailed study on the subject appears to be Wohnsiedler's work, published in 1960, in which it was shown that the alkyd resin not only exerts a plasticizing function but is also directly involved in the cross-linking of the system [9]. Wohnsiedler considered reactions between hydroxyl groups in the alkyd and hydroxymethyl and, in particular, alkoxymethyl groups in the

$$\text{alkyd} - \text{OH} + \text{RO} - \text{CH}_2 \text{N} -\!\!\big\langle\!\big\langle \quad \longrightarrow \quad \text{alkyd} - \text{O} - \text{CH}_2 \text{N} -\!\!\big\langle\!\big\langle \quad + \text{ROH}$$

melamine resin as the most important in cocondensation; however, several other possibilities were also indicated. The self-condensation of melamine resins leads to the formation of methylene linkages between the aminotriazine structures with a simultaneous loss of alcohol and formaldehyde, as was shown by IR spectroscopy.

Wohnsiedler's work has been the basis for more thorough investigations of the curing process. Due to their relative structural simplicity, HMMM resins have been the melamine resins most widely used in these experiments.

The view of transetherification being the predominant curing reaction with hydroxyl/carboxyl-containing polymers has been supported by other authors [10,11], however, dissenting views have also been expressed [12]. Today, we do have a relatively good knowledge about the importance of the various structures present in melamine resins with regard to both self-condensation of the amino resin and cocondensation with hydroxyl-containing polymers.

The systems are extremely complex, however, and a total understanding of the process is still lacking. A special difficulty encountered with melamine resins is that the main curing reactions lead to the same type of structures, i.e., N—CH_2—N and N—CH_2—O—C, already present in the resins before curing. A proper monitoring of the curing reactions by spectroscopic methods is, therefore, extremely difficult, and a comprehensive analysis of this type does not seem to have been reported.

It is generally recognized that the curing of melamine resin systems is extremely pH-dependent. Both acids and bases can be used to catalyze the process, and it is known that the catalyst accelerates not only the reactions leading to cocondensation, but also self-condensation. In practice, the main use of alkaline catalysis is in connection with nonetherified melamine resins. These are used as cross-linking agents for some applications, e.g., in the corrugated board industry. Starch is then normally employed as the main polymer, and the amino resin is added in order to improve the water resistance of the glue joint. Both cocondensation and self-condensation of the melamine resin seem to take place. The main reactions involved are shown in Figure 5.3.

Nonetherified melamine resins are of no importance for paints and lacquers. As mentioned above, methylated or butylated resins are normally used for this purpose. These resins are nonreactive at high pH, but very fast-curing in the presence of acid. An acid catalyst, usually pTSA (p-toluene sulphonic acid), is, therefore, often added to the system. However, the increase in curing rate is sometimes obtained at the expense of certain film properties.

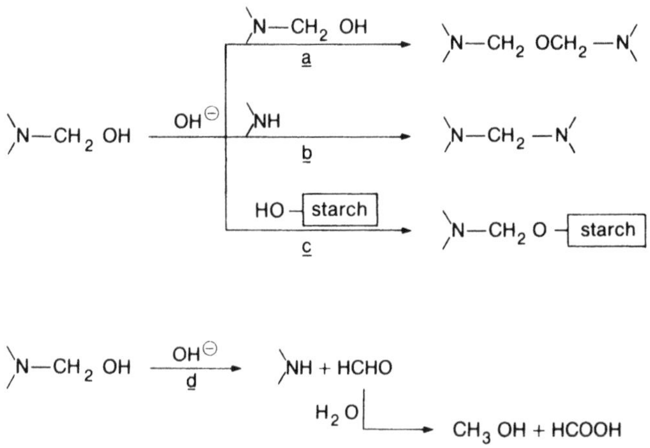

Figure 5.3 Curing reactions of starch-melamine resin systems under alkaline conditions. (From Ref. 13.)

The catalyst used is normally a strong acid, such as a sulphonic acid, sulphuric acid, or boron trifluoride. Hydrochloric acid should not be used since it may react with formaldehyde generated in the curing process to form the highly toxic bis(chloromethyl) ether:

$$HCl + HCHO \longrightarrow ClCH_2OH \xrightarrow{H^{\oplus}} ClCH_2OCH_2Cl$$

Even if the curing process is performed without the addition of an acid, it is reasonable to regard the condensation reactions to be acid-catalyzed since the alkyd molecules always contain free carboxyl groups. Only with model alkyds, in which the acid value is brought down towards zero by using especially effective esterification catalysts, or with other types of hydroxyl-containing polymers, such as polyetherpolyols or certain polyacrylates, does a catalyst-free curing exist. Hence, the alkyd-melamine resin curing is in practice always accelerated by some kind of acid present, and a division may then be made between strong and weak acid catalysis.

5.2.1 Cocondensation

It has been shown that the effect of an acid catalyst is different for resins having a high concentration of mono(alkoxymethyl)-amino groups, i.e.,

$$\rangle\!\rangle - N \overset{\textstyle CH_2OR}{\underset{\textstyle H}{<}}$$

and those having a high concentration of di(alkoxymethyl)amino groups:

$$\rangle\!\rangle - N \overset{\textstyle CH_2OR}{\underset{\textstyle CH_2OR}{<}}$$

Using the two model compounds HMMM and TMMM (trimeth-oxymethyl melamine) for kinetic studies, Berge et al. found that in moderately acid solution a partially methylolated resin decomposes much faster than one lacking NH groups [14]. At very low pH, on the other hand, the relative rates of decomposition are reversed, HMMM being much more reactive than TMMM. It was found that the decomposition of HMMM and TMMM followed the kinetics of specific and general acid catalysis, respectively.

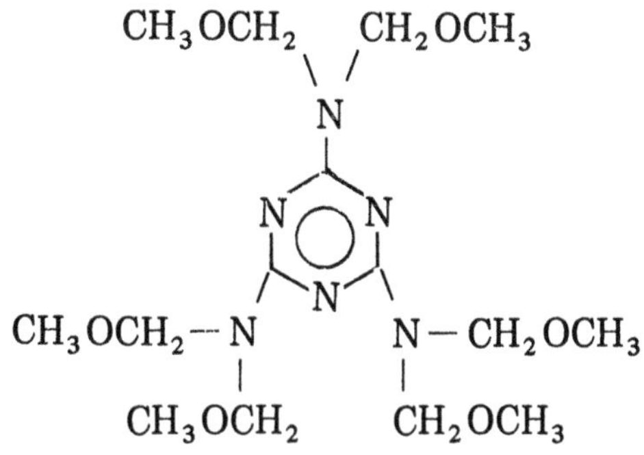

$$CH_3OCH_2 \quad CH_2OCH_3$$
$$\diagdown \diagup$$
$$N$$

$$\underset{}{N} \overset{}{\diagup} \underset{N}{\overset{}{\bigcirc}} \overset{}{\diagdown} N$$

$$CH_3OCH_2 - N \qquad N - CH_2OCH_3$$
$$| \qquad\qquad\quad |$$
$$CH_3OCH_2 \qquad CH_2OCH_3$$

HMMM

$$H \quad CH_2OCH_3$$
$$\diagdown \diagup$$
$$N$$

$$N \diagup \bigcirc \diagdown N$$

$$CH_3OCH_2 - N \qquad N - CH_2OCH_3$$
$$| \qquad\qquad\quad |$$
$$H \qquad\qquad H$$

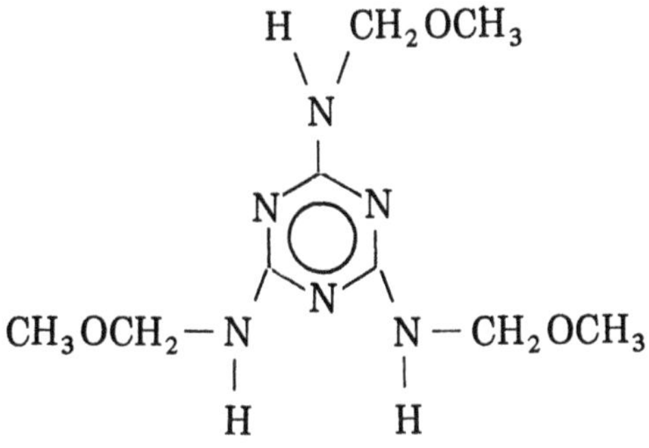

TMMM

The difference in decomposition mechanisms of the two melamine resin types is important since it suggests a difference also in the curing mechanisms [7]. Different behaviors of the two resins in the curing with alkyds have also been observed. Fully methylolated and etherified resins (such as HMMM) need strong acids as catalysts. Partially methylolated resins, on the other hand, also cure readily with weaker and undissociated acids.

(These resins also need a curing catalyst, however. In completely neutral, salt-free systems the curing process is extremely sluggish.)

It was later pointed out by Blank that curing by general acid catalysis is not only restricted to those resins having a high concentration of secondary amino groups, but also includes those containing hydroxymethylamino groups, since these will decompose under acidic conditions, eliminating formaldehyde and giving rise to NH groups [15,16]:

$$\text{)>}-\text{N}\Big\langle\begin{matrix}\text{CH}_2\text{OR}\\ \text{CH}_2\text{OH}\end{matrix} \quad \xrightarrow{\text{H}^{\oplus}} \quad \text{)>}-\text{N}\Big\langle\begin{matrix}\text{CH}_2\text{OR}\\ \text{H}\end{matrix} \quad + \text{ HCHO}$$

The curing of the fully methylolated and etherified resins is considered to involve an initial, rapid protonation of the ether oxygen, followed by a slower substitution reaction, as shown in Figure 5.4, path a.

The need for a considerable amount of strong acid as catalyst is evident from Table 5.3 [17]. At low catalyst level resin M1, which has the highest concentration of hydroxymethyl-amino groups, is the fastest curing, and resin M3, having the highest concentration of methoxymethylamino groups, is the slowest. At catalyst levels of $\geqslant 1\%$, calculated on solid amino resin, the relative rates of curing are completely reversed. Since the curing behavior of the resin systems at low levels of pTSA is the same as for melamine resins only (see Table 5.5), it seems that self-condensation of the melamine resin predominates over co-condensation with the hydroxyl-containing polymer below a certain amount of acid catalyst. In the gel time experiments presented in Table 5.3, a polyether polyol was used instead of a poly-ester polyol, i.e., an alkyd, in order to avoid the introduction of

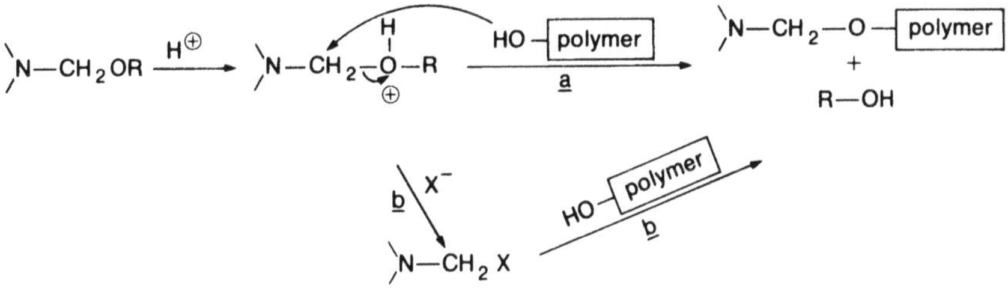

Figure 5.4 Cocondensation of a melamine resin with a hydroxyl-functional polymer under acidic conditions. R is preferably alkyl, but may also be H. X is Br or I.

Table 5.3 Gel Time (min) of a 3:1 Combination of a Polyol, P,[a] and Melamine Resins[b] M1-M4, at 130°C.

Catalyst (% pTSA)[c]	Resin			
	P/M1	P/M2	P/M3	P/M4
0	>60	>60	>60	>60
0.25	19	25	55	28
0.50	12	10	11	10
1.0	11	9	6	8
2.0	10	8	6	9

[a]P is a polyether polyol having a hydroxyl number of 120 mg KOH/g.
[b]The ratio melamine/formaldehyde/methanol of the melamine resins are: M1 1/5.50/3.00, M2 1/5.50/4.75, M3 1/5.50/5.50 and M4 1/4.75/4.75. All melamine resins had an initial pH of 8.0-8.2.
[c]The amount of pTSA as catalyst is calculated as solids on solids on amino resin.
Source: Ref. 17.

extra carboxyl groups into the system. Most probably, the same relative reactivities would have been obtained with an alkyd as the hydroxyl-containing polymer.

Different views have been advanced as to whether the co-condensation of a melamine resin and a hydroxyl-containing polymer is a S_N1 reaction, proceeding via the intermediate carbonium ion N—CH_2^+, or a S_N2 reaction, as illustrated in Figure 5.4. Most experimental data seem to support the bimolecular substitution mechanism. The experiments presented in Table 54., in which the effect of addition of Br⁻ and I⁻ to the reaction mixture has been investigated, is a further support of the S_N2 mechanism [17,18].

S_N2 reactions in polar solvents are known to be catalyzed by the addition of iodide and bromide ions. The catalytic effect is due to the fact that these ions combine a high nucleophilicity with a good leaving group ability. An enhancement of the reaction mechanism. The experiments presented in Table 5.4, in

Table 5.4 Gel Time (min) of a 3:1 Combination of Polyol P and Melamine Resin M2 in the Presence of 0.5% pTSA and 0 or 6% NaBr or NaI[a]

	Curing Temperature	
Salt	110°C	130°C
—	35	10
NaBr	20	8
NaI	12	7

[a]Calculated as solids on solids on amino resin. P and M2 are the same as in Table 5.3.
Source: Ref. 17.

cation of the S_N2 mechanism. The reactions involved are illustrated by path b of Figure 5.4.

Addition of iodide ions to the binder system results in strong discoloration, presumably due to oxidation of I^- to I_2. The bromide addition, however, seems not to cause any severe negative effects and could be of practical applicability. Preliminary tests using NaBr as a catalyst for the curing of a short oil alkyd-melamine resin system have given promising results. The reactivity of the mixture is increased without much reduction of pot life.

As mentioned above, melamine resins having a high concentration of secondary amino groups cure according to the mechanism of general acid catalysis. Accordingly, these systems show a very good cure response with weak acids, such as alkyl acid pyrophosphate. In fact, a comparison between this catalyst and pTSA as to the reaction between a hydroxyl-containing acrylic resin and a NH-rich melamine resin clearly showed that the stronger acid is a poorer catalyst [15].

Figure 5.5 shows the reaction pathway for the curing involving secondary amino groups. The initial protonation may occur not only on the ether oxygen but on the amino nitrogen atom as well. However, O-protonation is a prerequisite of the β-elimination which could proceed according to either an E1 or an E2 mechanism.

The formation of imino groups is reversible, and addition of an alcohol may re-form an alkoxymethylamino group. Since the curing is usually performed at a temperature above the boiling point of the alcohol of the melamine resin (normally methanol or butanol), addition would either take place to hydroxyl groups of the polymer or to nucleophilic groups of another

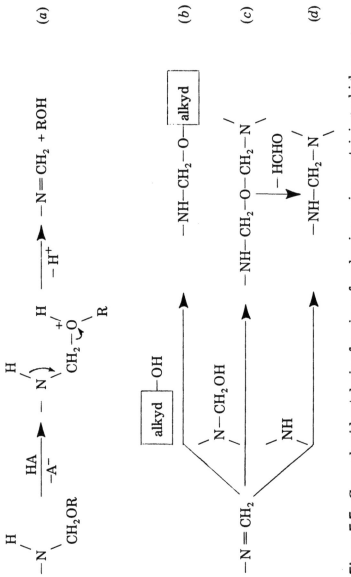

Figure 5.5 General acid catalysis of curing of melamine resins containing a high concentration of secondary amino groups. Pathway b leads to cocondensation, pathways c and d to self-condensation.

melamine resin molecule. The reactions, leading to cocondensation of the two polymers and self-curing of the melamine resin, respectively, are illustrated in Figure 5.5.

It should be pointed out that even if spectroscopic data strongly supports the mechanism of curing via imino groups, there are no indications of the quantitative importance of this route compared to other curing reactions. As mentioned before, monitoring the normal type of curing process, i.e., the formation of N—CH$_2$—O and N—CH$_2$—N linkages, is difficult since these structures are already abundant in the resins before curing. In most coating systems one can, of course, expect an overlap of specific and general acid catalysis. It is reasonable to believe that the reaction mechanism of Figure 5.4, path a, plays a role also in the curing of melamine resins having unusually high NHCH$_2$OR content.

The same type of curing mechanism, involving a Schiff base intermediate, has been suggested for low temperature curing of melamine resins using very high amounts of catalyst [19]. The type of melamine resin capable of undergoing curing at temperatures below 100°C is of the same type as that described above, i.e., resins having high concentrations of secondary amino groups. The catalyst used in this case is a strong acid, e.g., pTSA, however.

It has also been proposed that the Schiff base intermediate participates in a completely different type of curing reaction, as is shown in Figure 5.6 [20]. The imino group acts as dienophile in a Diels-Alder reaction, in which a conjugated double bond system, originating from the oil, serves as diene. There is no experimental evidence for this type of curing reaction. However, imine bonds are generally known to function as dienophiles in Diels-Alder reactions [21]. Imines, such as the one in Figure 5.6, which contain an electron-attracting substituent, would be expected to be especially reactive towards most dienes due to lowering of the energy of the lowest unoccupied orbital [22].

Figure 5.6 A Diels-Alder reaction between a Schiff's base, formed
by acid-catalyzed decomposition of a melamine resin, and a diene.

Conjugated double bonds are present to a certain degree in
most drying oils and fatty acids. Dehydrated castor oil (DCO),
which is frequently employed in alkyds used for curing with melamine resins, contains 25-30% conjugation. In addition, isolated
olefinic bonds present in linoleic and linolenic acid may isomerize
to conjugated systems under acidic conditions.

Conjugated systems in fatty acid residues are known to be
reactive dienes in Diels-Alder reactions if the olefinic groups are
in a cisoid conformation [23].

Trans trans substituted dienes may easily satisfy this condition and are consequently active towards dienophiles [24,25].
In *cis trans* and especially in *cis cis* substituted dienes, on the
other hand, the formation of the cisoid conformation is sterically
hindered through nonbonded interaction, as illustrated below.

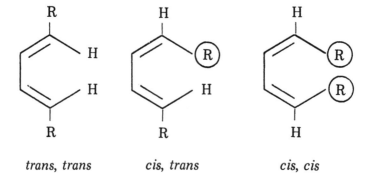

trans, trans   cis, trans   cis, cis

The number of *trans trans* unsaturated systems is small in most naturally occurring oils and fatty acids. However, *cis* to *trans* isomerization is a well-known phenomenon in the chemistry of fatty acids (elaidization), and it may take place both in the preparation of the alkyd and during the acid-catalyzed curing process. Most alkyds based on drying oils, therefore, contain a certain number of conjugated olefinic groups prone to act as dienes in Diels-Alder reactions.

The Diels-Alder reaction would not be expected to give any major contribution to the cross-linking process. However, the proposed reaction nicely explains why, at the same alkyd/melamine resin ratio, alkyds based on drying oils usually give harder films than those based on nondrying oils when cured with melamine resins under acidic conditions. Drying based on autoxidation of the fatty acid residues would be expected to be very slow in this case since no autoxidation catalyst is present.

5.2.2 Self-condensation

The alkyd resin could, in principle, undergo self-condensation due to autoxidation or intermolecular esterification or etherification. These reactions are slow, however, and are not regarded as important for the formation of the polymer network.

Self-condensation of amino resins, on the other hand, is a fast process in the presence of acid and must always be taken into consideration. The degree of methylolation and etherification of the melamine resin and the amount of acid catalyst used govern the ratio of cocondensation to self-condensation.

The self-condensation of a series of melamine resins has been studied in gel time tests and the results are shown in Table 5.5. These results, and other studies, demonstrate that

Table 5.5 Gel Time (min) of the Melamine Resins M1-M4 at 130°C[a]

Catalyst (% pTSA)	Resin			
	M1	M2	M3	M4
0	>60	>60	>60	>60
0.25	10	16	39	15
0.50	9	16	26	14
1.0	8	12	19	12
2.0	8	10	17	9

[a]M1-M4 are the same as in Table 5.3. The amount of catalyst is calculated as solids on solids.
Source: Ref. 17.

1. the self-curing of all resins is extremely slow in the absence of a catalyst;
2. highly etherified resins are relatively slow at all levels of added catalyst;
3. the partially etherified resins are reactive and need only small amounts of catalyst to reach full reactivity;
4. for the partially etherified melamine resins self-condensation competes favorably with cocondensation with alkyd resins, especially at low levels of catalyst. (Compare the gel times of Table 5.3 and Table 5.5.)

From the above study, as well as from other investigations [15], it appears that the rate of self-curing of the melamine resins in the presence of a strong acid catalyst is dependent on the degree of etherification but relatively independent of the degree of methylolation. This would indicate that pathway b of Figure 5.7 is the predominant self-curing mechanism. This bimolecular displacement reaction is analogous to the main cross-linking reaction

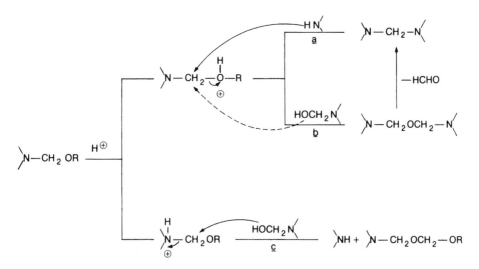

Figure 5.7   Self-curing of melamine resins under strongly acidic conditions. R is H or alkyl.

of the cocondensation process (Fig. 5.4, path a). However, this seems unlikely since the secondary amino group would be expected to be a stronger nucleophile than the hydroxyl group. Consequently, the reaction according to path a of Figure 5.7 should be much faster than the one according to path b .

  As pointed out earlier, hydroxymethylamino groups may decompose under acidic conditions to give secondary amino groups. A comparison of the curing behavior of melamine resins with that of a related compound, a glycoluril-formaldehyde resin, showed that the latter needed considerably more acid catalyst to cure properly [8]. Since glycoluril-formaldehyde resins have weak basic properties compared to melamine resins, a curing mechanism involving an initial N-protonation (pathway c of Fig. 5.7) would be less favorable. These findings and other results [17.18] suggest that self-curing of melamine resins under strongly acidic conditions occurs according to pathways c and a of Figure 5.7. The dimethylene diether, formed in path c, would be expected to gradually decompose with the emission of formaldehyde.

The self-condensation of the fully methylolated and etherified melamine resin must involve a diffusion of moisture into the film to give hydrolysis of alkoxyl groups [26]. The hydroxymethylamino groups formed will take part in self-curing reactions, as is discussed above (see Fig. 5.7).

The partially alkylated melamine resins may also undergo self-condensation according to general acid catalysis. This is analogous with cocondensation of these resins with hydroxyl-containing polymers using weak acid catalysts (see p. 155). Figure 5.5 illustrates this curing process. Addition of a hydroxy-methylamino group (pathway c) or a secondary amino group (pathway d) to the unstable Schiff base intermediate leads to self-curing.

5.3 CURING OF HIGH SOLIDS SYSTEMS

As mentioned earlier, high solids alkyd resins are usually of low functionality and of relatively low molecular weight, thus requiring large levels of cross-linking agent. Since the curing reactions are predominantly condensation reactions, each cross-linkage formed will result in a weight loss. The large weight loss of high solids systems due to larger amounts of condensation products will, therefore, counteract some of the gain in application solids. Nevertheless, the total reduction in emission of volatile materials obtained with appropriately formulated high solids paints, as compared to conventional systems, is considerable.

The weight loss during cure is dependent upon the structure of the alkyd and the amino resin, as well as on the mechanism of curing. The type of alcohol used in the melamine resin prepara-

tion is of particular importance since the alcohol is always eliminated during cross-linking, regardless of the curing mechanism.
For instance, elimination of three molecules of butanol from a
(theoretical) hexabutoxymethyl melamine results in a weight loss
of 34%, calculated on amino resin only. The corresponding
figure for hexamethoxymethyl melamine (HMMM) is 24%.

Bis(alkoxymethyl)amino groups (I in Table 5.6) and
secondary amino groups (II in Table 5.6) both participate in cocondensation with elimination of alcohol only, although the
mechanisms of curing are different (see Section 5.2.1 and Figs.
5.4 and 5.5). Amino groups having both an alkoxymethyl and a
hydroxymethyl group (III in Table 5.6), on the other hand, may
also give elimination of formaldehyde. The situation is a bit more
complicated with the latter group, however, since the condensation reaction may also involve the hydroxymethyl group, in
which case curing occurs with elimination of water only, as shown
in Figure 5.8.

It has been found that a strong acid catalyst, giving curing
according to specific acid catalysis, favors pathway b of Figure
5.8, whereas a weak acid, which gives general acid catalysis, predominantly leads to curing according to path a [15,16]. In practice, it is likely that the two routes of curing take place in concert.
The weight loss of monomeric melamine resins containing the
three functional groups I, II, and III is given in Table 5.6. This
Table shows theoretical values of weight loss with respect to the
melamine resins only. Since the fraction of the molecule lost in
the curing process is much smaller for the alkyd than for the
amino resin, the weight loss in percent of the total weight of the
binder system is normally higher, the higher the degree of melamine resin self-condensation.

Table 5.6 Weight Loss in Percent Solid Resin in the Curing of
Trifunctional Melamine Resins Containing Various Functional
Groups[a]

Functional Group	$R = CH_3$	$R = C_4H_9$
I. $-N\big\langle\,^{CH_2OR}_{CH_2OR}$	24	34
II. $-N\big\langle\,^{CH_2OR}_{H}$	36	57
III. $-N\big\langle\,^{CH_2OR}_{CH_2OH}$	53 (a)	65 (a)
	15 (b)	11 (b)

[a](a) and (b) refer to the pathways of Figure 5.8.

In addition to the weight loss originating from reactions of
functional groups, one must always count on some loss of material
during curing due to evaporation of unreacted starting materials
and other volatile products. Such low molecular components are
usually more abundant in high solids than in conventional systems.

In conventional alkyd-melamine resin systems only a limited amount of the available reactive groups in the cross-linking
agent participates in the curing process. A decreased mobility of
the growing polymer network limits the actual functionality of a
melamine resin in such a system. A resin of HMMM type has been

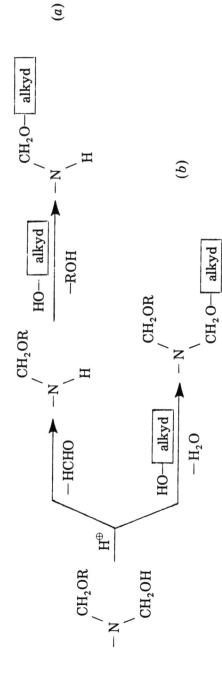

Figure 5.8   Acid-catalyzed curing of an alkyd resin with a partially etherified melamine resin.

found to possess an actual functionality of 2-3 per triazine ring, corresponding to participation of about half of the methoxy-methyl groups [6].

The alkyds used in combination with melamine resins for high solids formulations tend to be linear with hydroxyl groups at the ends. As a result of their low functionality these systems reach the gel point after a high degree of conversion, i.e., only after a high percentage of functional sites of the alkyd resin having reacted. This means that the actual functionality of the cross-linking agent in high solids systems needs to be higher than for conventional systems. A functionality of as high as 4 per triazine ring has been reported [6]. This is an important issue for high solids coatings since it limits the choice of melamine resins suitable as cross-linking agents. Partially methylolated or partially etherified resins, which are usually highly condensed, often do not contain enough reactive functional groups to yield a fully cured film. Therefore, highly methylolated and etherified products seem to be the melamine resins of choice for high solids systems. A schematic picture of the difference in network struc-ture between a high solids and a conventional system is given on the facing page. In conventional systems with a high functionality of the alkyd resin, adequate cross-linking takes place over a wide range of melamine resin level. With a higher melamine resin-to-alkyd ratio, the actual functionality of the amino resin decreases. The percentage of a HMMM resin in the total resin mixture is typically between 15 and 25.

High solids systems, on the other hand, require consider-ably higher levels of cross-linking agent and are much more sensi-tive to variations in the melamine resin-to-alkyd ratio. This is nicely illustrated by the development of film hardness at different HMMM levels for a conventional and a high solids system on bak-

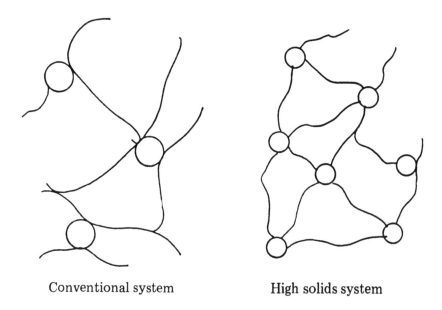

Conventional system   High solids system

ing (Fig. 5.9) [6]. The HMMM level in high solids systems is normally between 25 and 35% of the total resin.

Since the amino resins used in high solids systems normally have a very high amount of nonvolatile material, an increase in melamine resin-to-alkyd ratio may lead to a higher total solids content of the formulation. If the melamine resin level is increased above a certain value, the cross-linking density of the polymer network will be too low if the cross-linking is due to cocondensation reactions only. The reason for this is that the actual functionality of the melamine resin is reduced from the maximum value according to the formula (see also p. 58):

$$F_{actual} = \frac{F_{maximum}}{1 + n}$$

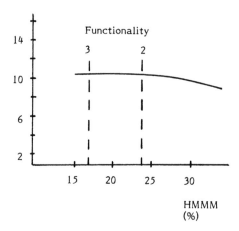

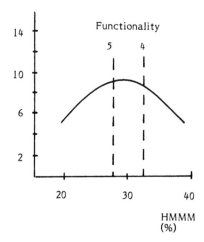

Figure 5.9 Film hardness as a function of hexamethoxymethyl melamine (HMMM) content in the curing of a (a) conventional and (b) high solids system at 125°C for 20 min. pTSA was used as catalyst in an amount of 0.4% calculated on amino resin. The approximate functionality of the melamine resins is indicated by the dotted lines. (From Ref. 6.)

where n is the fraction of reactive groups in the melamine resin present in excess of hydroxyl groups in the alkyd resin.

If, for example, an alkyd resin of molecular weight 840, having two hydroxyl groups per molecule, is cured with a pure HMMM (which has a molecular weight of 210 and six reactive methoxyl groups) using a strong acid catalyst and an alkyd to melamine resin ratio of 2:1, the ratio of reactive functional groups in the system will be:

$$\frac{\text{moles OH}}{\text{moles OCH}_3} = \frac{2}{1} \cdot \frac{210}{840} \cdot \frac{2}{6} = \frac{1}{6}$$

Thus, the fraction of OCH_3 groups in the HMMM resin present in excess of OH groups in the alkyd is

$$n = \frac{6 - 1}{1} = 5$$

Assuming only cocondensation, this gives for the amino resin:

$$F_{actual} = \frac{6}{1 + 5} = 1$$

Consequently, only one functional group per triazine moiety will react. Even if a pure HMMM is an unrealistic cross-linking agent, the calculation above clearly shows that cocondensation alone will not always suffice to give a three-dimensional network. Evidently, the use of such high levels of melamine resin requires a considerable amount of self-condensation in order to achieve an appropriate cross-linking density.

The self-curing that occurs with a fully methylolated and etherified melamine resin, such as HMMM, cannot be directly explained by the reaction pathways of Figure 5.7. A self-condensation of a resin of this type must be preceded by a hydrolysis with the elimination of methanol:

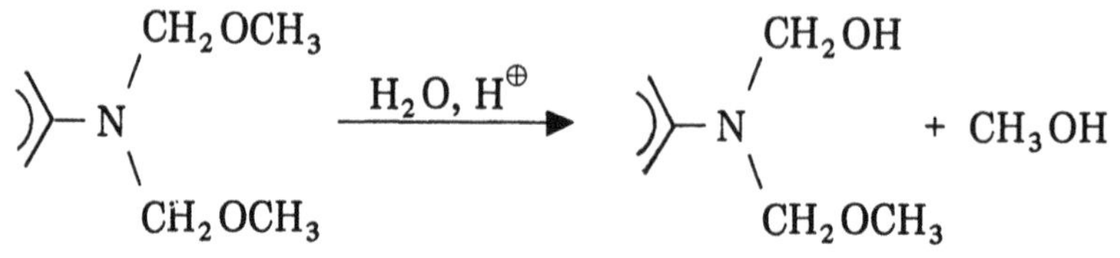

$$
\text{>>}-N
\begin{array}{c}
\diagup CH_2OCH_3 \\
\diagdown CH_2OCH_3
\end{array}
\quad
\xrightarrow{\;H_2O,\ H^{\oplus}\;}
\quad
\text{>>}-N
\begin{array}{c}
\diagup CH_2OH \\
\diagdown CH_2OCH_3
\end{array}
\quad + \ CH_3OH
$$

The emission of methanol according to the above reaction is experimentally verified [15]. The extent of this reaction must be dependent on the amount of water present in the formulation.

The cocondensation of HMMM-type melamine resins with hydroxyl-containing polymers has been discussed in detail in Section 5.2.1. The curing reaction taking place in the presence of a strong acid as catalyst is shown in Figure 5.4, path a.

5.4 ACID CATALYST FOR HIGH SOLIDS SYSTEMS

As has been pointed out in the two preceding sections, an acid catalyst is usually needed to get acceptable curing performance of high solids alkyd-melamine resin systems. The most common catalyst is p-toluene sulphonic acid (pTSA), frequently used in "blocked" form as its amine salt to ensure a satisfactory pot life of the resin mixture.

The use of an acid catalyst in high solids formulations often influences the film properties in a negative way, however. At moderate-to-high catalyst levels the resulting film tends to be water sensitive. Furthermore, both free acids and their salts impart a too high conductivity to the system to permit electrostatic spraying. Lastly, the use of an acid catalyst may result in poor adsorption to some pigments. In recent years, two different approaches have been taken to overcome these problems.

5.4.1 Hydrophobic Sulphonic Acids

A series of sulphonic acids with varying degree of hydrophobicity were evaluated with regard to film properties [27]. The resin system used was a high solids saturated polyester and a melamine resin of HMMM type (see p. 145). The acids included in the test are shown in Figure 5.10. Methane sulphonic acid (MSA), p-toluene sulphonic acid (pTSA), and dodecylbenzene sulphonic acid (DDBSA) are all completely water-soluble; dinonylnapthalene disulphonic acid (DNNDSA) has a limited solubility and dinonyl-naphthalene sulphonic acid (DNNSA) is insoluble in water. A catalyst level of 1% pTSA on total resin solids was employed, and the amounts of the other acids were adjusted so as to yield the same acidity in each case.

MSA was found to develop the best hardness, but the films were brittle and showed poor water resistance. The disulphonic acid, DNNDSA, gave film properties similar to those of pTSA, except for considerably improved water resistance. The other naphthalene sulphonic acid, DNNSA, also gave relatively good water resistance, but the cure and hardness of the film were poor. This hydrophobic acid was probably not properly dissolved and evenly distributed in the polar resin mixture, thus giving a low cross-link density of the film. DDBSA gave acceptable water resistance, but the film obtained showed an extremely poor adhesion to the steel substrate used. This has been explained as being due to the surface active properties of this acid, resulting in the formation of a monolayer of the sulphonate on the surface of the steel panel. The linear alkyl aryl sulphonate is considerably more surface active than naphthalene based sulphonic acids, the latter having bulky hydrophobic groups which render the formation of closed packed monolayers more difficult.

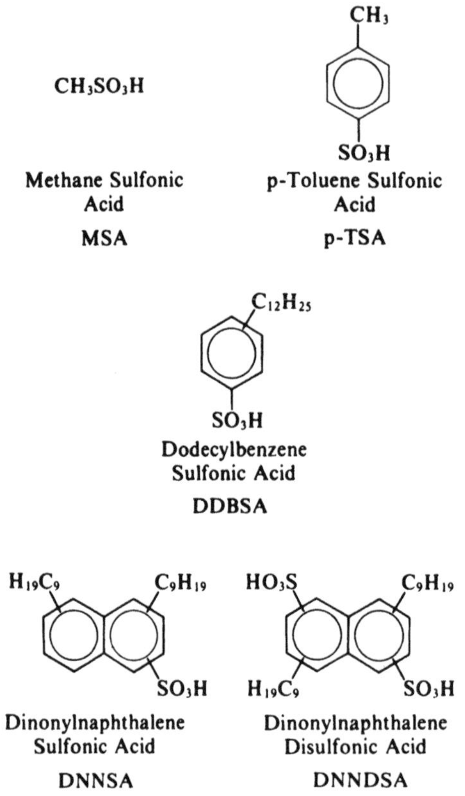

CH₃

CH₃SO₃H

SO₃H

Methane Sulfonic   p-Toluene Sulfonic
     Acid   Acid
     MSA   p-TSA

C₁₂H₂₅

SO₃H
Dodecylbenzene
 Sulfonic Acid
    DDBSA

H₁₉C₉   C₉H₁₉   HO₃S   C₉H₁₉

       SO₃H   H₁₉C₉   SO₃H
Dinonylnaphthalene   Dinonylnaphthalene
  Sulfonic Acid   Disulfonic Acid
     DNNSA   DNNDSA

Figure 5.10 Alkyl and alkylaryl sulphonic acids used as curing catalysts.

A further observation from this study is that the disulphonic acid, DNNDSA, seems to promote cocondensation of the alkyd and the melamine resin and suppress melamine resin self-condensation, as compared with pTSA. This effect is probably not related to differences in acid strength since the two aromatic sulphonic acids have approximately the same pKa, 0.6-0.7. (The second pKa of the disulphonic acid is, of course, considerably higher.) Instead a physical interpretation of the data is proposed.

The more hydrophobic acid, DNNDSA, would be expected to be more compatible than pTSA with the resin mixture,
especially with the less polar polyester component. Consequently,
pTSA would accumulate more at the polymer/metal interface and,
when in the polymer mixture, be more localized in the amino resin
than in the polyester microdomains. DNNDSA, on the other
hand, in particular in the form of the disodium salt, is known to
be an effective hydrotrope for promoting the solubility and coupling of two dissimilar phases. By promoting a better interaction
between the reactants during the critical stages of film formation
when solvent and reaction by-products are driven off, cocondensation re actions should be favored.

5.4.2 Oxime Esters as Latent Catalysts

Amine salts of sulphonic acids are established as a type of blocked
or latent acid catalyst. Although this approach has a positive effect on pot life, it does not solve the problems associated with the
presence of strong electrolytes (see p. 170). The use of latent
catalysts in which the acid and the blocking group are covalently
bound is a route that may overcome these problems. To this end,
a series of esters of sulphonic acids have been synthesized and evaluated as catalysts for curing of a high solids polyester-melamine
resin system [28]. The melamine resin used was of the HMMM
type, and an amount of catalyst corresponding to 0.6-0.7% p-
toluene sulphonic acid on total resin solids was employed.

During the curing process the sulphonic acid is generated
from the oxime sulphonic ester. The decomposition may follow
two different routes, as shown in Figure 5.11. Path a is analogous
to the well-known Beckmann rearrangement of oximes to substituted amides. Also oxime esters are known to undergo this
reaction. Route b is a fragmentation reaction sometimes referred
to as abnormal Beckmann rearrangement. The two routes prob-

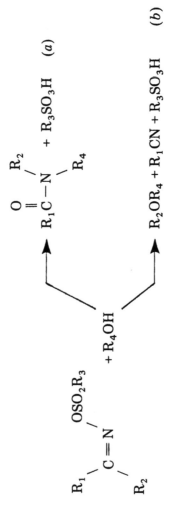

Figure 5.11 Thermolysis of oxime sulphonic esters.

ably proceed via a common intermediate [29], as shown in
Figure 5.12, although other mechanisms are also possible for
the Beckmann rearrangement [30]. The stability of the carbonium ion R_2^+ is probably decisive for the extent of the fragmentation reaction (route b).

It was found that the ratio of rearrangement products to
fragmentation products could be governed by the choice of R_1
and R_2. More importantly, the rate of dissociation of the oxime
ester varied considerably depending on the nature of the groups
R_1 and R_2. Electron-accepting groups generally had a retarding
effect on the reaction rate, probably through destabilization of
the intermediate carbonium ion.

By selecting the appropriate R_1 and R_2 groups, a catalyst
with tailor-made effect may be obtained. As always, a compromise between cure response and pot life has to be made. The following oxime esters were considered to be most suitable. (The
stereochemistry of the products was not established, however.)

pTSA was used as the acid component in all cases but one. Variation of the sulphonic acid along the lines described in Section
5.4.1 might lead to a catalyst which combines the low conductiv-

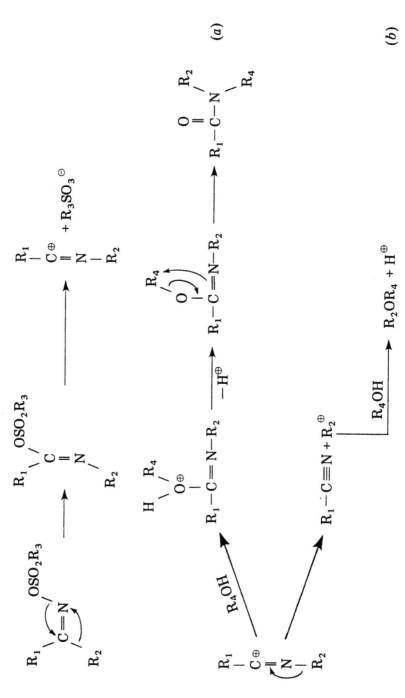

Figure 5.12 Mechanism of rearrangement and fragmentation of oxime sulphonic esters.

ity of the oxime esters with the good water resistance of the cured films obtained with the more hydrophobic acids.

REFERENCES

1. M. F. Kooistra, *J. Oil Col. Chem. Assoc.*, *62*:432 (1979).

2. J. P. Walsh, *Mod. Paint Coat.*, *73*:42 (Nov. 1983).

3. G. Christensen, *Prog. Org. Coat.*, *8*:211 (1980).

4. B. Tomita and S. Hatono, *J. Polym. Sci.*, *Polym. Chem. Ed.*, *16*:2509 (1978).

5. M. G. Lazzara, *J. Coat. Technol.*, *56*:19 (1984).

6. N. J. Albrecht and W. J. Blank, *Adv. Org. Coat. Sci. Technol. Ser.*, *4*:1 (1982).

7. A. Berge, *Proc. 3rd Intern. Conf. Org. Coat. Sci. Technol.*, Athens, Greece, 1977, p. 31.

8. J. O. Santer, *Prog. Org. Coat.*, *12*:309 (1984).

9. H. P. Wohnsiedler, *ACS Div. Org. Coat. Plast. Chem.*, *20/2*:53 (1960).

10. J. Dörffel and U. Biethan, *Farbe Lack*, *82*:1017 (1976).

11. R. Seidler and H. J. Graetz, *Proc. FATIPEC Congr.*, Verlag Chemie, Weinheim, West Germany, 1962, p. 282.

12. S. E. Stromberg, *ACS Div. Org. Coat. Plast. Chem.*, *29/2*:321 (1969).

13. K. Holmberg, *Polym. Bull.*, *6*:553 (1982).

14. A. Berge, B. Kvaeven, and J. Ugelstad, *Europ. Polym. J.*, *6*:981 (1970).

15. W. J. Blank, *J. Coat. Technol.*, *51*:61 (1979).

16. W. J. Blank, *J. Coat. Technol.*, *54*:26 (1982).

17. K. Holmberg, *Org. Coat. Sci. Technol.*, *8*:125 (1986).

18. K. Holmberg, *J. Oil Col. Chem. Assoc.*, *61*:359 (1978).

19. K. Holmberg, *Polym. Bull.*, *11*:81 (1984).

20. K. Holmberg, *J. Oil Col. Chem. Assoc.*, *61*:356 (1978).

21. T. L. Gilchrist and R. C. Storr, *Organic Reactions and Orbital Symmetry*, Cambridge University Press, London, 1972, p. 92.

22. K. Fukui, in *Molecular Orbitals in Chemistry, Physics and Biology* (P.-O. Löwdin and B. Pullman, eds.), Academic Press, New York, 1964, p. 513.

23. W. Carruthers, *Some Modern Methods of Organic Synthesis*, Cambridge University Press, London, 1971, p. 131.

24. W. G. Bickford, J. S. Hoffman, D. C. Heinzelman, and S. P. Fore, *J. Org. Chem.*, *22*:1080 (1957).

25. H. M. Teeter, J. L. O'Donnel, W. J. Schneider, L. E. Gast, and M. J. Danzig, *J. Org. Chem.*, *22*:512 (1957).

26. W. J. Blank and W. L. Hensley, *J. Paint Technol.*, *46*:46 (1974).

27. L. J. Calbo, *J. Coat. Technol.*, *52*:75 (1980).

28. W. J. Mijs, W. J. Muizebelt, and J. B. Reesink, *J. Coat. Technol.*, *55*:45 (1983).

29. J. March, *Advanced Organic Chemistry*, 3rd Ed., Wiley-Interscience, New York, 1984, p. 932.

30. R. M. Palmere, R. T. Conley, and J. L. Rabinowitz, *J. Org. Chem.*, *37*:4095 (1972).

6
The Reactive Diluent Concept

One attractive way of achieving high solids alkyd paints is to use a reactive diluent which can function as a solvent in the formulation of the coating and which during curing is converted into an integral part of the film. The curing may be initiated by raising the temperature or by adding a catalyst. In the latter case the cure process should preferably take place at ambient or force-dry temperatures.

6.1 LINSEED OIL

Linseed oil is the traditional reactive diluent. Together with long oil alkyds it has been used as such and, more importantly, as the thermally polymerized stand oil or the oxidatively polymerized boiled (or blown) oil. The cure process involves autoxidation, i.e., reaction with air oxygen. The curing is always performed in the presence of an autoxidation catalyst, a so-called drier, which is

normally a combination of metal salts. These systems may be regarded as two-component systems although they are normally
not thought of as such. The major component is the alkyd-linseed oil mixture having the metal salt catalyst dissolved in it.
The other component is oxygen and the reaction does not start
until the air has come in contact with the binder-diluent component.

Linseed oil, although still in use, is not important as a
reactive diluent any longer. Its viscosity is too high and the curing rate with air-drying alkyds is considered too low for modern
use. Since it has a long history of being the reactive diluent of
choice and, more important, since it is the parent compound of
a family of newer reactive diluents, its curing behavior justifies
some attention.

The mechanism of curing of linseed oil-alkyd systems is
extremely complex. It is well known that conjugated fatty acid
esters cure much faster than unconjugated ones [1] and that,
among the unconjugated derivatives, linoleic and linolenic acid
esters are more than two orders of magnitude faster than esters
of oleic acid [2]. Since both drying alkyds and linseed oil (polymerized or not) contain multiunsaturated fatty acid chains, it is
reasonable to believe that the curing of these systems is analogous
to the curing of pure linoleates and linolenates, both of which
have been investigated in some detail.

The first stage of the curing of linoleates is probably formation of hydroperoxides via a highly delocalized allylic radical,
as is shown in Figure 6.1 [3,4].

An alternative mechanism has been proposed for the formation of hydroperoxides involving a cyclic complex between
oxygen and an allylic structure of the linoleate. The reaction is
shown in Figure 6.2 [5].

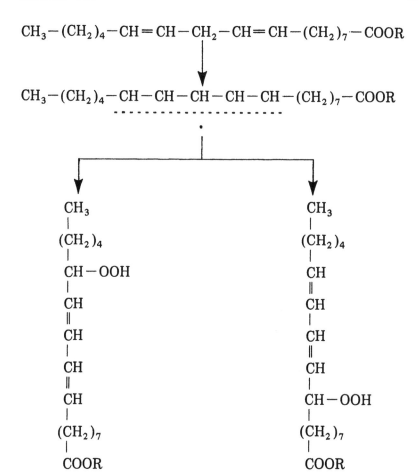

Figure 6.1 Formation of hydroperoxides from linoleates via a delocalized allylic radical.

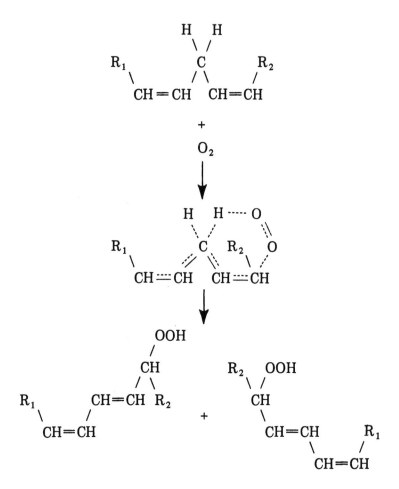

Figure 6.2 Formation of hydroperoxides from linoleates via a cyclic transition state.

Both reaction mechanisms lead to the same type of hydroperoxide in positions allylic to a conjugated diene system. It is
clearly established that hydroperoxides of this type are the major
reaction products in the initial stage of the autoxidation [6].
More highly oxidized products, such as dihydroperoxides [7] (Fig.
6.3) and Diels-Alder adducts between oxygen and the diene system [1,8] (Fig. 6.4), may also form. The latter reaction will most
easily occur when the diene system is *trans trans*, this being the
most favored configuration of a diene in the Diels-Alder reaction
[9]. The original configurations of the double bonds in linoleates
are *cis cis*. *Cis-trans* isomerization is, however, likely to take place
in the formation of hydroperoxides according to the reaction
pathways of both Figure 6.1 and Figure 6.2.

The second stage of the curing of linoleates involves decomposition of the hydroperoxides with formation of cross-linked
structures. A schematic representation of such a structure for the
system air-drying alkyd-linseed oil is shown in Figure 6.5.

The cross-linking is believed to be a free radical reaction,
as is shown below [1]. The terminating reactions, which are also
those leading to cross-linking, involves formation of carbon-carbon bonds, as well as ether and peroxide bonds.

$$ROOH \rightarrow RO\cdot + \cdot OH$$

$$RO\cdot + R'H \rightarrow ROH + R'\cdot$$

$$R'H + \cdot OH \rightarrow R'\cdot + H_2O$$

$$R'\cdot + R'\cdot \rightarrow R' - R'$$

$$RO\cdot + R'\cdot \rightarrow R\!-\!O\!-\!R'$$

$$RO\cdot + RO\cdot \rightarrow R\!-\!O\!-\!O\!-\!R$$

Apart from the reactions originating from decomposition of hydroperoxides, free radicals may also be formed by dissociation of

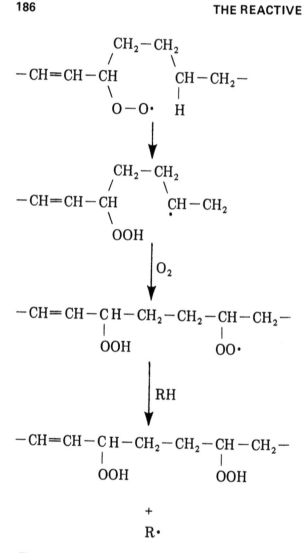

Figure 6.3   Formation of dihydroperoxides.

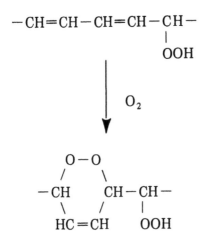

Figure 6.4 Diels-Alder addition of oxygen to linoleate hydroperoxide.

the cyclic peroxide of Figure 6.4. The allylic radical formed may initiate a free radical chain polymerization, as shown in Figure 6.6.

As appears from the above discussion, the linkage between fatty acid residues of the alkyd and the reactive diluent—linseed oil—may be a carbon-carbon bond, an ether bond, or a peroxide bond. A quantitative estimate of the ratio between the first and the two latter types of bonds can be obtained by simply measuring the amount of oxygen built into the film during curing. From reactions with model systems it is known that a high curing temperature favors the formation of C—C bonds [10].

The main function of the autoxidation catalysts seems to be to assist in the decomposition of the hydroperoxide [11]. Cobalt and manganese ions seem to be particularly effective in this respect. As Me(II) ions they are known to participate in a one electron transfer reaction leading to alkoxy radicals. In doing this the metal ion is oxidized and must, consequently, be reduced back by another reaction, which may involve decomposition of

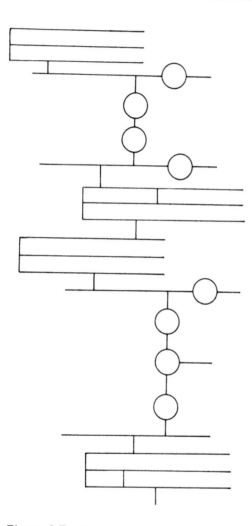

Figure 6.5 Schematic structure of a cross-linked film of air-drying alkyd-linseed oil.

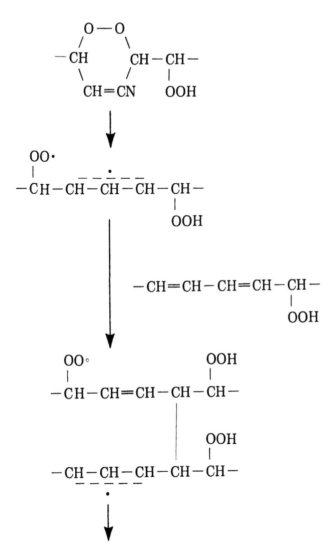

Figure 6.6 Cross-linking of conjugated systems starting from decomposition of a cyclic peroxide.

another hydroperoxide molecule. The reactions involved for the cobalt ion in a suggested redox system are shown below:

$$R\!-\!OOH + Co^{2+} \rightarrow RO\cdot + OH^- + Co^{3+}$$

$$R\!-\!OOH + Co^{3+} \rightarrow R\!-\!OO\cdot + Co^{2+} + H^+$$

The role of the autoxidation catalyst is more complex than that, however. In practice the cobalt and/or manganese salt is usually combined with so-called auxiliary driers, which do not catalyze autoxidation when used alone, but which seem to enhance the action of the true catalyst. The reader is referred to Solomon [1] for a more comprehensive discussion of the function of driers.

6.2 OTHER MULTIUNSATURATED COMPOUNDS

Linseed oil in its various forms suffers from severe drawbacks as a reactive diluent. In its natural form it is too slow-curing for most applications. In its polymerized forms (see p. 181) the curing rate is acceptable but now the viscosity is too high. A considerable amount of white spirit normally needs to be added in order to reach a suitable application viscosity.

A systematic search has been made to find multiunsaturated compounds which, better than the vegetable oils, combine a good reactivity in autoxidation with a low viscosity [12]. Naturally, the compounds should also be nonvolatile and have a good compatibility with air-drying alkyds.

A range of compounds of various structures and with varying degrees of unsaturation were investigated. Those containing conjugated double bonds were found to be faster curing and to absorb less oxygen during the curing process than those containing isolated double bonds. This is in accordance with results from earlier studies on fatty acid esters, where it was shown that conjugated structures gave more C—C cross-linking, whereas compounds containing 1,3-diene structures (as most drying oils do) mainly cross-linked by C—O—C or even C—O—O—C bridges [13].

The compound in the series that best fulfilled the criteria regarding curing rate, viscosity, volatility, and odor was the acetal 1 of Figure 6.7 containing both conjugated and isolated double bonds. As is shown in Figure 6.7, the compound, although seemingly complex in structure, is made from simple and inexpensive starting materials. The yields of the two reaction steps are reported to be above 70 and 90%, respectively. Compared to the intermediate alcohol 2, the acetal 1 was much faster in "set to touch," had a higher boiling point and less odor, but a slightly higher viscosity. The physical properties of the acetal are shown in Table 6.1.

The drying time of the acetal alone lies in the same region as most air-drying alkyds. The temperature dependence of the curing rate of the product with a Co—Pb drier combination added is shown in Figure 6.8.

The drying process was monitored by IR spectroscopy and by refractive index measurements. By comparison with corresponding data for alkyd drying it is proposed that the two curing reactions basically proceed by the same mechanism.

Pigmented paints were formulated using the acetal as reactive diluent for a conventional air-drying alkyd. With a pigment-to-resin ratio of 1:0.98, a formulation with 95% solids content by weight could be made. Without the reactive diluent the solids content at the same viscosity was below 70%. Various ratios of alkyd to reactive diluent were tested, and it was found that the higher the ratio, the better the drying properties. The conventional alkyd formulation had the shortest "set to touch" time.

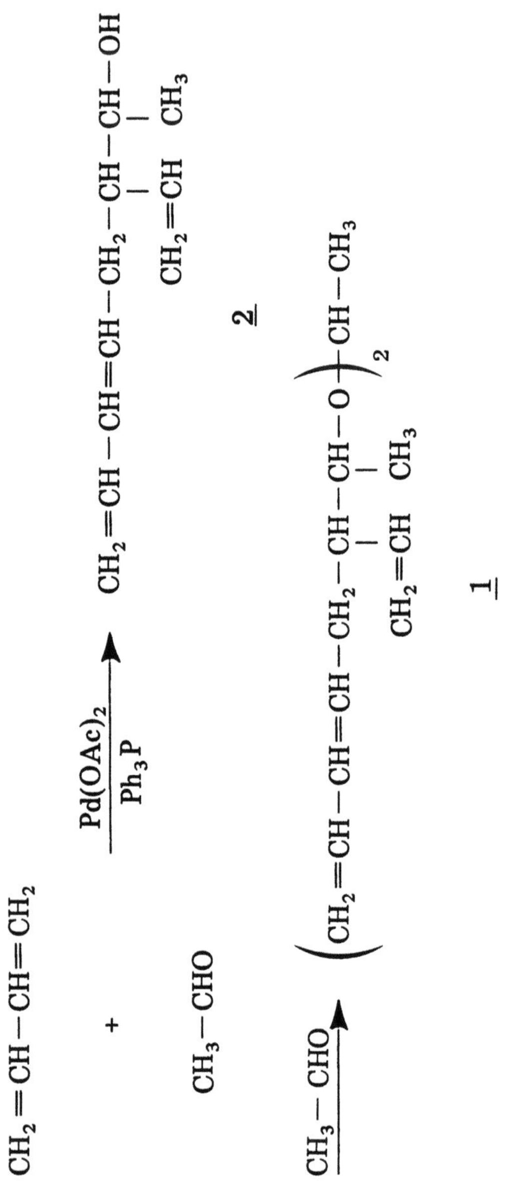

$$CH_2=CH-CH=CH_2$$

$$+$$

$$CH_3-CHO$$

$$\xrightarrow[\text{Ph}_3\text{P}]{\text{Pd(OAc)}_2}$$

$$CH_2=CH-CH=CH-CH_2-CH-CH-OH$$
$$\underset{CH_2=CH}{|}\quad\underset{CH_3}{|}$$

$$\underline{\textbf{2}}$$

$$CH_3-CHO$$

$$\left(CH_2=CH-CH=CH-CH_2-CH-CH-O\right)_2 CH-CH_3$$
$$\underset{CH_2=CH}{|}\quad\underset{CH_3}{|}$$

$$\underline{\textbf{1}}$$

Figure 6.7 Preparation of a multiunsaturated acetal suitable as a reactive diluent.

Table 6.1 Physical Properties of Acetal 1[a]

Boiling point, °C/mm Hg	315-320/760
	144-148/2
Viscosity (25°C), mPa · s	8.63
Refractive index (25°C)	1.4908
Density (15°C), kg/m^3	902.5
Compatibility with long oil alkyds	good to very good
Compatibility with short oil alkyds	fair to poor

[a]See Figure 6.7 MW 330.
Source: Ref. 12.

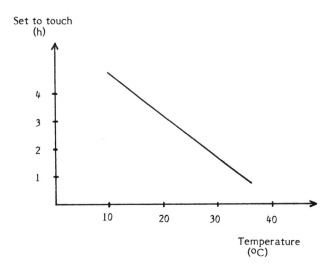

Figure 6.8 Drying time, "set to touch," vs. temperature of acetal 1 of Figure 6.6. A Co—Pb drier combination is used.

Evidently in this case a compromise would have to be made between drying time and solids content of the paint.

6.3 ALLYL ETHERS

Allyl ethers were found long ago to undergo metal-catalyzed oxidation with subsequent copolymerization with fatty acid derivatives, thus bringing about drying of alkyd paints [14]. The allyl
ethers are most conveniently prepared by reaction of alcohols or
glycols with acrolein. The monoallyl acetal of the tetrafunctional
alcohol pentaerythritol gives a glycol with built-in air drying
properties. The diallyl acetal of the same tetrol gives a compound
which is both air-drying and polymerizable through the olefinic
bonds [15].

$$CH_2=CH-CH \begin{array}{c} O-CH_2 \\ \diagup \quad \diagdown \\ \diagdown \quad \diagup \\ O-CH_2 \end{array} C \begin{array}{c} CH_2-OH \\ \diagup \\ \diagdown \\ CH_2-OH \end{array}$$

Monoallyl acetal of pentaerythritol

$$CH_2=CH-CH \begin{array}{c} O-CH_2 \\ \diagup \quad \diagdown \\ \diagdown \quad \diagup \\ O-CH_2 \end{array} C \begin{array}{c} CH_2-O \\ \diagup \quad \diagdown \\ \diagdown \quad \diagup \\ CH_2-O \end{array} CH-CH=CH_2$$

Diallyl acetal of pentaerythritol

The most interesting type of products along these lines seems to
be the vinyl dioxolanes, prepared by reaction of acrolein with
vicinal glycols [16].

$$
\begin{array}{c}
\text{R}-\text{CH}-\text{CH}_2 \\
\mid \qquad \mid \\
\text{OH} \quad \text{OH}
\end{array}
\;+\;
\begin{array}{c}
\text{CHO} \\
\mid \\
\text{CH} \\
\parallel \\
\text{CH}_2
\end{array}
\;\rightarrow\;
\begin{array}{c}
\text{R}-\text{CH}-\text{CH}_2 \\
\mid \qquad \mid \\
\text{O} \qquad \text{O} \\
\diagdown \; \diagup \\
\text{CH} \\
\mid \\
\text{CH} \\
\parallel \\
\text{CH}_2
\end{array}
$$

The allylic CH group of the vinyl dioxolanes shows a considerable reactivity towards oxygen. The acetal structure evidently makes the compound more susceptible to oxidative attack than normal allyl ethers. Furthermore, the ring structure seems to promote the autoxidation rate, the corresponding acyclic compound being less reactive with oxygen. This indicates that opening of the slightly strained dioxolane ring system might be a driving force in the autoxidation reaction.

Compounds of special interest appear when the vicinal glycol used as starting material for the vinyl dioxolane contains other reactive functional groups as well. The compound can then be incorporated into low molecular prepolymers or even be reacted into the alkyd molecules during curing at elevated temperatures. In this case the curing of the alkyd may be truly enhanced by the presence of the reactive diluent since the vinyl dioxolane derivative of the polymer may not only participate in the normal type of autoxidation via hydroperoxides but may also polymerize through acrylate structures formed, as shown in Figure 6.9.

The formation of the acrylate structure in the drying of vinyl dioxolane compounds has been confirmed by spectroscopic studies [17,18].

196

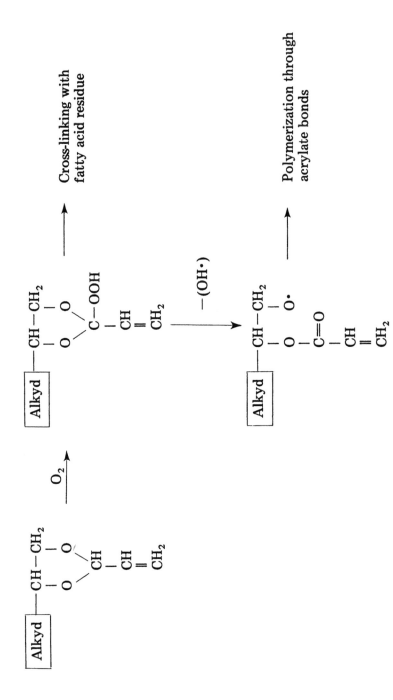

Figure 6.9 Drying of alkyds modified by vinyl dioxolanes.

6.4 ACRYLATE ESTERS

Esters of acrylic acid and methacrylic acid should constitute an
attractive route to reactive diluents since the chemistry of the
free radical polymerization of these monomers is so well explored.
These compounds, however, are generally considered unsuitable
because of less favorable toxicity profiles and because they react
with the oxime stabilizers, which are used to give stable one-pack-
age systems. Pentaerythritol triacrylate [19], as well as higher
alkyl methacrylates, such as stearyl and lauryl methacrylate [20],
are, however, reported to give stable systems. The latter compounds, all the same, seem less attractive since they give a poor
surface cure, yielding soft films. The triacrylate of pentaerythritol, on the other hand, is especially recommended for use with
acrylated alkyds in order to get a good curing.

In a search to find acrylic monomers especially designed
for reactive diluent use, dicyclopentenyloxyethyl methacrylate
(DPOMA) was found to be the most promising candidate [21].

DPOMA

DPOMA is an autoxidizable methacrylate that can be cured at
ambient temperature by oxidatively initiated polymerization. The
related compound dicyclopentyl methacrylate can also be cured
with oxygen using a cobalt catalyst [22]. Unlike DPOMA, how-

ever, this somewhat smaller molecule is relatively volatile and has a penetrating and atrocious odor.

DPOMA contains both a polymerizable methacrylate double bond and an allylic group, which can serve as a free radical source in the presence of oxygen. It is, therefore, capable of both homopolymerization to form a polyalkylmethacrylate and copolymerization with coating vehicles containing unsaturated groups. Drying oil alkyds are examples of such vehicles and the copolymerization is catalyzed by conventional metal driers.

Monomeric DPOMA is stable in the absence of an autoxidation catalyst. In the presence of driers, such as cobalt salts, it polymerizes in bulk in a few days. Addition of antioxidants of the oxime type is a way to retard the autoxidation and to achieve extended package stability. In thin films, the volatile oxime evaporates and normal cure is obtained.

Oximes of aldehydes and ketones are believed to function by forming coordination complexes with the cobalt ion, thereby preventing it from partitioning in redox type reactions [23].

$$\mathrm{Co} \left[\begin{array}{c} R \\ \diagdown \\ C = NOH \\ \diagup \\ R^1 \end{array} \right]_6^{3+}$$

The physical properties of DPOMA are given in Table 6.2.

Furthermore, DPOMA is reported [21] to be "practically nontoxic" in both oral and dermal tests ($LD_{50} > 5$ g/kg), to be nonmutagenic in Ames tests, and to be nonsensitizing in skin sensitization test.

Table 6.2 Physical Properties of Dicyclopentenyloxyethyl
Methacrylate (DPOMA)[a]

Boiling point, °C	350
Viscosity (25°C), mPa · s	15-19
Refractive index (22°C)	1.496
Density (25°C), kg/m^3	1064
Flash Point (Pensky-Martens CC), °C	> 93
Solubility parameter, cal$^{1/2}$/cm$^{3/2}$	8.6
Compatibility with drying alkyds	very good

[a]MW 262.
Source: Ref. 21.

The autoxidation of DPOMA is believed to be initiated by
a free radical attack at the allylic position, generating a chain reaction as illustrated in Figure 6.10. The hydroperoxides formed are
decomposed by one-electron transfer reactions with the cobalt
catalyst (see p. 187).

The radicals formed by the autoxidation process may generate homopolymerization of DPOMA through the methacrylate
functionality. In thin films with good contact between oxygen
and the monomer, curing at ambient temperature results in a
methacrylate/oxygen-alternating polymer. (Oxygen has a much
higher reactivity than alkyl methacrylates towards carbon free
radicals.) In the presence of a drying alkyd a variety of curing reactions are likely to occur in concert, some of which are schematically illustrated in Figure 6.11. The alkyd not only polymerizes
with itself (route a), or with the reactive diluent (route b), but
also acts as an oxygen scavenger, thus suppressing the cocondensation between the methacrylate and oxygen (route e).

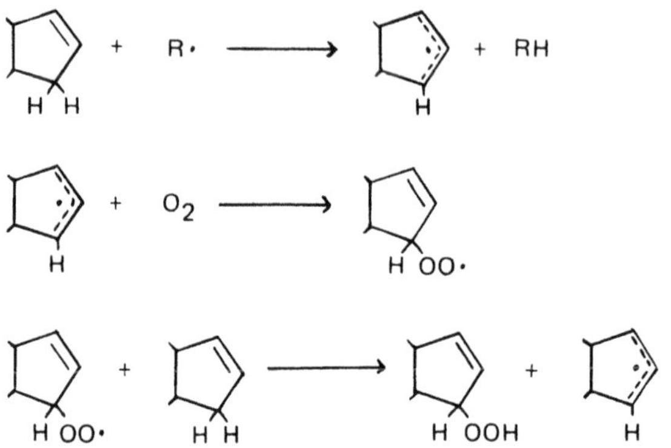

Figure 6.10 Autoxidation of dicyclopentenyloxyethyl methacrylate (DPOMA).

The cocondensation between the alkyd resin and the reactive diluent has been confirmed by analytical studies [21]. This reaction is believed to be one of the reasons for the good through-dry properties of these systems. Especially with medium and long oil alkyds the use of DPOMA as reactive solvent brings about a considerable improvement of the drying characteristics. Best results have been obtained with a ratio of DPOMA to alkyd of between 10:90 and 25:75. As is seen in Figure 6.12, early dry or set time is extended when the reactive diluent is used, whereas the through-dry time is reduced. Film properties are also reported to be less satisfactory with short oil alkyds. The reason for the poorer performance of DPOMA-short oil alkyd systems may be that the oxygen-scavenging effect of these resins is insufficient to prevent extensive formation of the methacrylate/oxygen copolymer.

The cocondensation of an alkyd and an acrylate-based reactive solvent can be favored by incorporating acrylate or methacrylate structures into the alkyd. This approach has been evaluated for the system acrylated alkyd-pentaerythritol triacry-

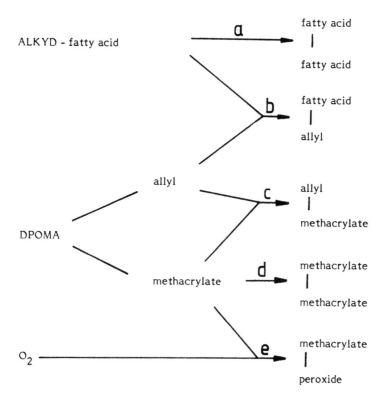

Figure 6.11 Curing reactions of the system drying oil alkyd-dicyclopentenyloxyethyl methacrylate (DPOMA) in the presence of air and an autoxidation catalyst.

late (PETA) [19]. The alkyd is simply prepared by esterifying a resin having a higher than usual hydroxyl number with acrylic acid using an acid catalyst.

   PETA is highly reactive towards acrylated resins and, being trifunctional, is an effective cross-linker. Like DPOMA-based systems, a volatile antioxidant needs to be added to the formulation in order to obtain a good package stability. PETA is probably not as good a solvent for drying alkyds as, for instance, DPOMA, and substantial amounts of normal solvents have to be

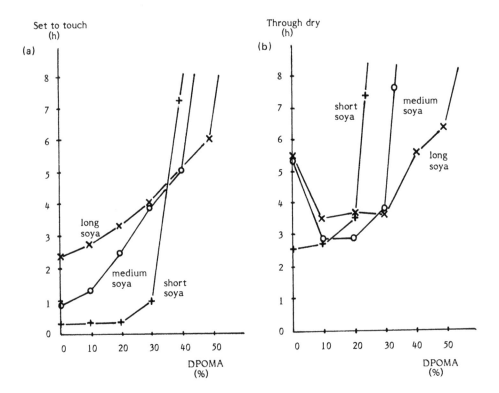

Figure 6.12 (a) Set to touch and (b) through-dry time vs. level of dicyclopentenyloxyethyl methacrylate (DPOMA) in three different soya oil alkyds with varying oil lengths. The systems were catalyzed by 0.06% Co and 0.06% Ca. (From Ref. 21.)

included in the formulation. The percentage of PETA on weight of solid resin normally lies between 20 and 40. In order to improve pigment dispersion a small amount of nonacrylated alkyd may also be added.

The first step in the curing of PETA-based systems is the normal autoxidation of fatty acid residues in the alkyd giving rise to hydroperoxide groups. Under the influence of metal driers decomposition of the hydroperoxides take place (see p.

$$CH_2O - C - CH = CH_2$$

with structure:

$$
\begin{array}{c}
\quad\quad\quad\quad O \\
\quad\quad\quad\quad \| \\
CH_2O -\!\!- C - CH = CH_2 \\
| \\
\quad\quad\quad\quad\quad O \\
\quad\quad\quad\quad\quad \| \\
HOH_2C - C - CH_2O - C - CH = CH_2 \\
| \\
\quad\quad\quad O \\
\quad\quad\quad \| \\
CH_2O - C - CH = CH_2
\end{array}
$$

PETA

187) with formation of free radicals. These initiate polymerization
of the acrylate groups leading to both homo- and cocondensation
reactions. Normal autoxidative drying of the alkyd takes place
simultaneously.

As seen in Table 6.3, PETA has a melting point around
ambient temperature. This is not important, however, since a normal solvent must also be added to the formulation in order to reduce the viscosity.

6.5 UNSATURATED POLYESTERS

From a pure chemical point of view, unsaturated polyesters as
a class do not fall within the definition of alkyd resins given on
p. 18. From a broader viewpoint, however, the unsaturated
polyester-styrene system must be regarded as closely related to
high solids alkyds. A chapter on the reactive diluent concept
would be incomplete unless a section dealing with this type of
system was included.

Unsaturated polyesters are produced in enormous quantities and they are considered one of the three principal thermoset

Table 6.3 Physical Properties of Pentaerythritol Triacrylate (PETA)[a]

Melting point, °C	25
Viscosity (25°C), mPa · s	600-800
Density (25°C), kg/m^3	1180
Flash point, °C	> 82

[a]MW 298.

plastic materials of economic importance. The resins can be cast, laminated, gel-coated, or molded. They are normally compounded with fibers, in particular fiber glass, or fillers, or both in the liquid stage and are then cured to yield thermoset articles. The plastics area of unsaturated polyesters is not within the scope of this monograph, however. The reader is referred to the relevant litera-ture [24-26] for a presentation of this topic.

The main application area of relevance to paints and lac-quers is gel coat. A gel coat is a polyester resin paint which nor-mally is sprayed onto a mold. It is used primarily in the marine industry as the exterior paint layer for various types of boats and also in the bathroom-fixture industry in the preparation of shower stalls and bathtub enclosures.

Furthermore, unsaturated polyesters have found use in highly durable, high gloss lacquers for wood. The use of these systems, however, has been severely restricted by the toxicity of styrene. The concern pertains both to worker exposure and to emission into the environment. Styrene is known to be an irri-tant to the eyes and mucous membranes and is a central nervous system depressant [27].

6.5.1 The System

"Unsaturated polyester" is a vague expression defined on p. 18. In general, the resin consists of a predominantly linear, unsaturated backbone polymer, thinned in styrene. An "activator" is also present. Immediately prior to use an organic peroxide or hydroperoxide is added. Free radicals are then generated and polymerization is initiated by attackof the free radicals on the styrene or polyester unsaturation. The structure and final properties of the resulting network depend on the relative amounts of the two kinds of unsaturation.

The curing temperature is governed by the choice of catalyst system. Ambient temperature is sometimes employed but force dry conditions seem to be most common. Radiation curing, especially by UV light, has found use for wood lacquers [28].

In kinetic studies low molecular fumarate esters are normally used as model compounds for the unsaturated polyesters. Since the reactivity ratios of both fumaric ester and styrene are below one, copolymerization is favored. The optimal mole fraction of diethyl fumarate in combination with styrene calculated from the copolymer equation is around 0.40 [29 This corresponds well with the empirically found optimum of styrene vs. fumarate unsaturation. This optimum can probably be accounted for by the maximum in cross-linking density occurring at this ratio between the polymerizable groups [30]. At the end of the reaction the conversion of styrene monomers was found to be greater than that of polyester unsaturations [31].

The copolymerization reaction normally presents an induction period; then the rate increases, passes through a maximum, and finally decreases. The induction period is related to the consumption of radicals by inhibitors present in commercial formulations. Recent kinetic studies indicate that different curing mechanisms operate, depending on the temperature. Whereas at lower temperatures first-order kinetics apply, at temperatures above 100°C the reactions are second-order with respect to unsaturated groups. This behavior has been explained to be related to differences in the initiator decomposition rate, giving rise to different types of termination steps of the polymerization [32].

The cured film will consist of oligomeric styrene chains cross-linking the polyester molecules. No true homopolymer of styrene is formed [33]. The degree of cross-linking is governed by the concentration of unsaturated groups in the resin backbone. An unsaturated polyester, therefore, normally contains saturated as well as unsaturated ingredients. The cured polyester-styrene system can be presented as follows:

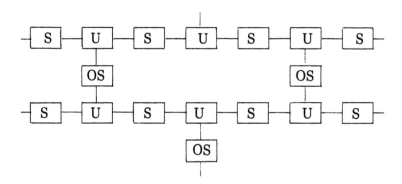

where

        S = saturated component of the polyester resin
        U = unsaturated component of the polyester resin
     OS = oligomeric styrene

Incorporation of air-drying functions in the unsaturated polyesters leads to products which in the interior of the coatings cure by normal vinyl polymerization and on the surface undergo autoxidation [34]. This is one of the approaches taken to overcome the inhibition exerted by oxygen on the vinyl polymerization (see Section 6.5.5).

6.5.2 The Polyester

Unsaturated polyester resins are in principle prepared by the same synthetic methods as alkyd resins (see Chapter 3). The molecular weight distribution curve of a typical unsaturated polyester was found to be in good agreement with that expected from Flory's theory of polycondensation kinetics [35].

Three types of raw materials are commonly used in the synthesis:

1. aliphatic, unsaturated dibasic acids
2. aromatic or saturated aliphatic di- or tribasic acids
3. aliphatic polyols

In addition, smaller quantities of monofunctional acids or alcohols are sometimes employed. These reactants are often post-added in order to bring down the hydroxyl number or the acid value, respectively, of the polyester.

The unsaturated dibasic acids may be regarded as the most critical raw material, as they impart the ability to copolymerize with the reactive diluent. Maleic anhydride and fumaric acid are usually employed and, because of economic considerations, maleic anhydride is used in the majority of commercial syntheses. During the resin preparation *cis-trans* isomerization of maleate to fumarate takes place and most unsaturated polyesters, therefore, contain predominantly *trans* olefinic bonds. The ratio of *trans* to *cis* double bonds in the resin is extremely important in order to achieve the right balance of properties of the cured product. As discussed in the next section, the reactivity ratios of maleate and styrene are very different from those of fumarate and styrene, and a high amount of fumaric ester is needed to get substantial cocondensation. As a result, high fumarate polyesters normally give harder and less flexible coatings than high maleate ones.

The *cis-trans* isomerization can conveniently be monitored by NMR spectroscopy; signals at δ 6.33 and 6.74 ppm correspond to *cis* and *trans* H, respectively [36]. The conversion to fumarate is very slow in neutral systems; if free carboxyl groups are present or if an acid catalyst is added, the isomerization is fast, however, and virtually complete formation of the *trans* isomer is possible [37,38]. Furthermore, the choice of raw materials used has been found to influence the degree of isomerization. Vicinal glycols give more *trans* isomer than nonvicinal [39], and ortho-phthalic acid gives more than isophthalic acid [40].

An important side reaction in the synthesis of unsaturated polyesters is the addition of alcohol to the double bond of maleate or fumarate. This reaction may account for as much as up to 20% double bond saturation and may, thus, strongly influence the mechanical properties of the styrene cross-linked final network [41]. The reaction has been found to be catalyzed by acids and favored by high temperatures. Alkyl-substituted ethylene glycols, such as 1,2-propylene glycol, add across the double bond with particular ease, whereas ethylene glycol and other diols having two primary hydroxyl groups need more forcing conditions [42].

The addition reaction is believed to proceed via a cationic intermediate. A suggested mechanism is shown in Figure 6.13 [38]. The succinic acid derivative formed may undergo internal cyclization giving a lactone [43]. (This reaction is probably only important for vicinal diols, i.e., when R' in Figure 6.13 is a two-carbon residue.)

The addition of alcohol to the olefinic bonds of the growing polyester backbone results in the formation of side chains and in an uncontrolled modification of the stoichiometry of the reac-

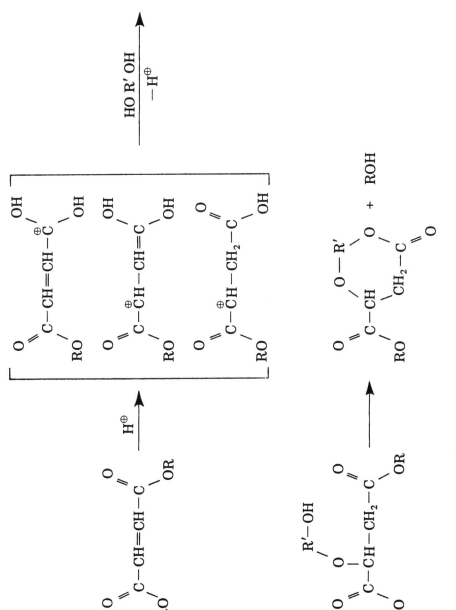

Figure 6.13 Addition of a glycol to fumarate or maleate.

tion. The only ways to avoid this side reaction seem to be either to work at very low temperature to to do the polyesterification in the absence of free acid. In principle, the former approach is possible by using acid chlorides and the latter by employing methyl esters of maleic or fumaric acid and performing transesterification with glycols.

Aconitic acid, itaconic anhydride, and mesaconic anhydride are also used to some extent to introduce backbone unsaturation. Economic considerations limit their use, however.

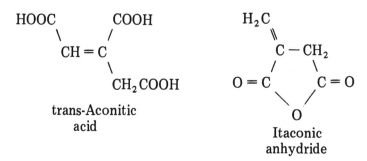

trans-Aconitic
acid

Itaconic
anhydride

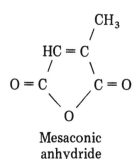

Mesaconic
anhydride

As mentioned earlier, saturated aliphatic, as well as aromatic, di- and tribasic acids are used in conjunction with maleic anhydride and fumaric acid to modify the chemical and mechanical properties of the resin. Adipic acid and phthalic anhydride are the most common examples of these two classes of acids. Generally, aromatic acids promote hardness and aliphatic acids promote

product flexibility. Since orthophthalic acid has a tendency to
form cyclic compounds of low molecular weight, substitution of
phthalic anhydride by isophthalic acid gives higher molecular
weight resins with a considerably higher melting point. Normally,
good chemical resistance and excellent physical properties of the
polyester are achieved with isophthalic acid.

A variety of other aliphatic dibasic acids, such as succinic,
azelaic, and sebacic acid, as well as dimer fatty acids, are used as
raw materials for unsaturated polyesters. In general, the longer
the distance between the carboxyl groups, the higher the flexibility of the cured resin film.

The glycols used in the resin syntheses are mostly low
molecular weight, aliphatic ones. Epoxides, such as epichlorohyd-
rin and cyclohexene oxide have also been used as the polyol component [44,45]. In this type of compound the oxirane ring serves
as a reactive vicinal glycol (see Section 3.4).

6.5.3 The Reactive Diluent

Styrene is by far the most common reactive diluent for unsaturated polyesters. Alkyl substituents in the styrene aromatic ring
may improve properties such as water resistance of the cured film
[46]. Other vinyl or acrylic monomers have also found some
use, usually however as co-monomers with styrene. The main advantage with styrene, apart from low cost, is good solvency and
high reactivity with fumarate unsaturation. Volatility and toxicity
are the main drawbacks.

The monomer reactivity ratios for styrene and diethyl
fumarate, as well as for styrene and diethyl maleate, are given in
Table 6.4. It can be seen that the fumarate favors copolymer formation, whereas the maleate gives more homopolymerization of

Table 6.4 Monomer Reactivity Ratios (at 60°C)

M_1	r_1	M_2	r_2	Ref.
Styrene	0.30	Diethyl fumarate	0.070	47
Styrene	6.5	Diethyl maleate	0.005	47
Vinyl acetate	0.031	Diethyl fumarate	0.32	48
Vinyl acetate	0.031	Di-n-hexyl fumarate	0.27	48
Vinyl acetate	0.17	Diethyl maleate	0.043	47
Methyl methacrylate	20	Diethyl maleate	0	49
Methyl methacrylate	17.5	Poly(ethylene glycol) fumarate	0.35	49

styrene. This difference in reactivity ratios of the two systems is the main reason why so much attention is being paid to the isomerization of maleate to fumarate.

The reactivity ratios of a few other monomers in combination with diethyl fumarate are also given in Table 6.4. The reactivity ratios for diethyl fumarate and di-n-hexyl fumarate in combination with vinyl acetate are almost identical, implying that the diethyl ester is a good model for dialkyl fumarates in general.

As can be seen from the table, vinyl acetate, similar to styrene, gives predominantly copolymerization with dialkyl fumarate [48]. Methyl methacrylate, on the other hand, gives mainly acrylate homopolymerization [49]. It must be remembered, however, that in the later stages of the curing process the immobility of the functional groups will be expected to lead to variations from the theoretical copolymerization scheme. Therefore, in using the copolymer equation to calculate the film composition, it is necessary to use modified values for the reactivity ratios.

Combinations of styrene and methacrylic esters have received attention as reactive solvent systems [50-53]. Optimizing the solvent blend in order to arrive at specific film properties is an alternative to optimization of the polyester resin. Particularly complex systems arise when multifunctional methacrylates, such as triethylene glycol dimethacrylate [53], are employed.

$$\underset{\begin{array}{c}| \\ H_2C = C - C - O -(\!CH_2CH_2O\!)_3C - C = CH_2 \end{array}}{\overset{\begin{array}{cc} H_3C & O \\ | & \| \end{array} \qquad \qquad \begin{array}{cc} O & CH_3 \\ \| & | \end{array}}{}}$$

Triethylene glycol dimethacrylate

A system of particular interest arises when two monomers are combined which give little mutual copolymerization and where only one of the components is able to copolymerize with the fumarate groups of the polyester. In this case two more or less independent networks will form, which is analogous to the formation of a so-called interpenetrating polymer network (IPN) by simultaneous polymerization of a monomer mixture through two distinct types of polymerization [54].

The combination dialkyl fumarate-methacrylic ester of a polyol-vinyl ester constitutes a system of this type. The reactivity ratios of a model system of this kind are shown on Figure 6.14. Substantial copolymerization will only take place between the fumarate and the vinyl ester.

An unsaturated polyester was cured using a mixture of vinyl versatate and ethylene glycol dimethacrylate as reactive diluent [29]. The vinyl ester cross-linking yielded a dense network, internally plasticized by the side chains: the homopolymer of vinyl versatate has a T_g of $-3°C$. The methacrylate only copolymerized with the polyester to a small extent; instead, a dense and stiff network of methacrylate homopolymer was formed. Thus,

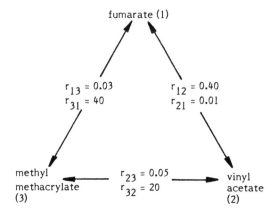

Figure 6.14 Reactivity ratios for copolymerization of diethyl fumarate, vinyl acetate, and methyl methacrylate. (From Ref. 55.)

an IPN-like structure, consisting of one hard and one more flexible polymer, was obtained.

6.5.4 The Catalyst System

Unsaturated polyesters are cured using a redox system usually consisting of an organic peroxide or hydroperoxide (the "initiator") in combination with an "activator," or "accelerator," which is normally either a cobalt salt or an aromatic amine. Typical combinations are benzoyl peroxide and N,N-dimethylaniline, or a hydroperoxide and a cobaltous salt of a medium chain organic acid, e.g., octoate [56,57].

The hydroperoxide-cobalt salt system is believed to generate free radicals by the same mechanism as was discussed for autoxidation catalysts (p. 187). Only trace amounts of cobalt salt are needed.

The redox system dialkylaniline-diacyl peroxide gives a very fast curing. The mechanism appears to involve generation of free radicals from both the benzoyl and the amine species, as shown in Figure 6.15.

The nitrogen radical cation formed (1 of Fig. 6.16) is highly stabilized due to delocalization into the phenyl ring. Consequently, this type of amine gives considerably lower initiation energy for curing than the corresponding trialkyl amines. The rate of cure can be governed by the choice of substituent on the aromatic ring. Electron-withdrawing groups, such as nitro or halogen, in the *para* position decrease the speed, whereas electron-donating substituents, e.g., alkoxyl, increase it [59].

Sometimes a complex system consisting of a hydroperoxide, a metal salt, and an amine is used to initiate polymerization. A multiligand complex between the metal ion and the amine is likely to form, and studies on one such system indicate that cobalt has a coordination number of 4 [60].

The function of the amine activator has been discussed by a number of authors and special attention has been payed to cobalt with 1,10-phenanthroline as ligand [1]. It seem probable that the main effect of the amine is related to a lowering of the energy required for reduction of Co(III) to Co(II), thus promoting the decomposition of the hydroperoxide. Other reducing agents, such a s tartaric or citric acid, are also employed for this purpose. The effect of these additives is two-fold. Apart from increasing the rate of cure, they also help preventing discoloration of the film. A high concentration of cobaltic salt often gives rise to an undesirable green color.

The organic peroxides and hydroperoxides most frequently used to initiate polyester curing are listed in Table 6.5. The commercial products are far from pure, however, but consist of

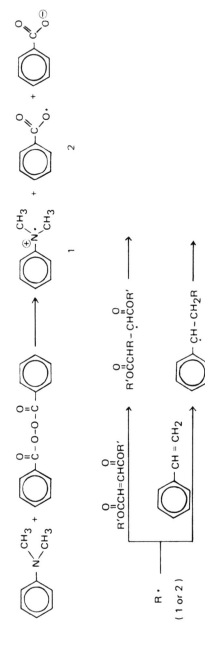

Figure 6.15 Curing of an unsaturated polyester-styrene system using benzoyl peroxide/N,N-dimethylaniline as initiator. (From Ref. 58.)

Table 6.5 Common Initiators for Low Temperature Curing of
Unsaturated Polyesters

Chemical Name	Structure

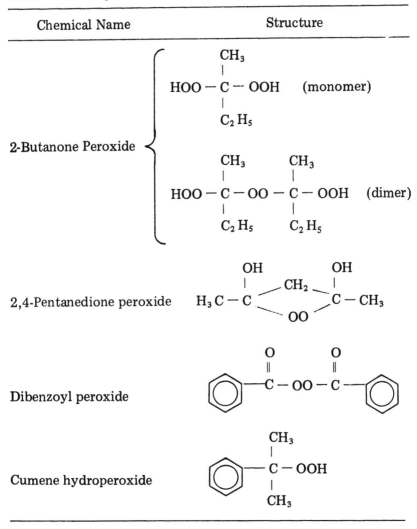

2-Butanone Peroxide

$$HOO - \overset{\overset{\displaystyle CH_3}{|}}{\underset{\underset{\displaystyle C_2H_5}{|}}{C}} - OOH \quad \text{(monomer)}$$

$$HOO - \overset{\overset{\displaystyle CH_3}{|}}{\underset{\underset{\displaystyle C_2H_5}{|}}{C}} - OO - \overset{\overset{\displaystyle CH_3}{|}}{\underset{\underset{\displaystyle C_2H_5}{|}}{C}} - OOH \quad \text{(dimer)}$$

2,4-Pentanedione peroxide

Dibenzoyl peroxide

Cumene hydroperoxide

a variety of structural isomers and mixtures of monomers and o oligomers. For the most widely used initiator, methyl ethyl ketone peroxide ("MEK-peroxide" or, correctly, 2-butanone peroxide), at least eight different species have been identified [1]. The monomer-to-dimer ratio of this compound has been found to be decisive for the reactivity of the system [61].

Aliphatic azo compounds, especially azonitriles, may also be used as initiators. A major advantage of these compounds compared to peroxides is that they are not subject to induced decomposition by various metal salt impurities. Their major drawback is that they cannot, like peroxides and hydroperoxides, be activated by special promotors at temperatures well below their normal thermal activation temperature. The use of the azo initiators is, therefore, restricted to higher temperature applications.

$$
\begin{array}{ccc}
R & & R \\
| & & | \\
R-C-N=N-C-R \\
| & & | \\
CN & & CN
\end{array}
$$

A symmetrical azonitrile

Recently, the curing of unsaturated polyester resins with reactive diluents has been performed using accelerators chemically incorporated into the polyester prepolymer. Tertiary amines of aryl or alkyl diethanolamine type were incorporated into the polyester during the resin preparation [62,63]. The procedure is outlined in Figure 6.16. Curing was performed with benzoyl peroxide as initiator. It was found that the amine accelerators were considerably more efficient when incorporated into the polymer chain than when used free. The pot lives, gel times, and cure times were significantly shorter when polymer-bound amines were used at all

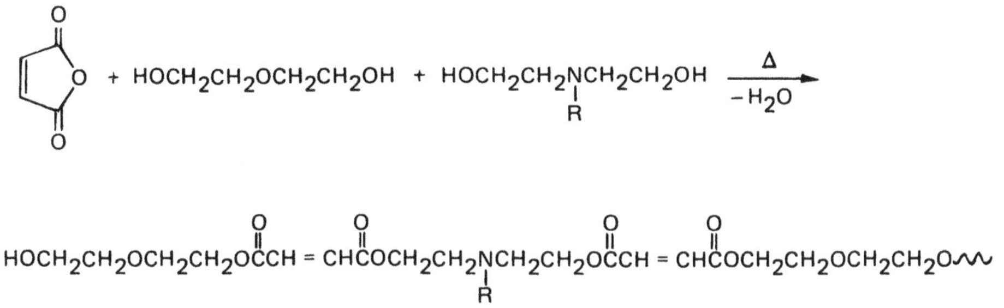

Figure 6.16 Preparation of an unsaturated polyester with built-in amine accelerator.

cure temperatures studied. Rate constants for initiation were obtained and the energy of initiation calculated using Arrhenius' plots. It was found that the initiation energies for the reactions with the polymer-bound accelerators were lower than for those using the corresponding free amines. It was also observed that less homopolymerization took place when polymer-bound accelerators were used. Higher cross-link densities and shorter polystyrene chains were formed. This is the expected behavior considering the initiating mechanism shown in Figure 6.15. The probability for copolymerization will be enhanced when one of the initiating radicals (*1* of Fig. 6.15) is already situated in the polyester backbone.

In another study a metal salt, as well as a tertiary amine, was incorporated into the polyester prior to cure [53]. The metal was introduced as the dimaleate. MEK peroxide was used as initiator.

$$(HOC-CH=CH-C-O^-)_2 Me^{2+}$$

Me = Co, Mn

The kinetics of the curing reactions were studied and compared with the corresponding processes using the accelerator and promotor in free form. A considerable reduction in initiation energies were obtained with the polymer-bound species.

Inhibitors are usually added to unsaturated polyester resins in order to prolong storage life, reduce discoloration, and retard the rate of cure to prevent crazing or cracking. The most widely used types of inhibitors are phenols and quinones, although the parent phenol is weak in activity. Hydroquinone, p-benzoquinone, and their mono- and disubstituted derivatives are very common [64].

Similar to initiators, combinations of inhibitors are often used to optimize a formulation. Various quarternary ammonium salts, such as benzyl-trimethylammonium halides, are often used together with the phenol-type inhibitors. The system of inhibitors must always be balanced against the initiating system; a high level of inhibitor may give too long an induction period of the curing process.

6.5.5 Air Inhibition

The curing of unsaturated polyester-styrene systems is strongly inhibited by air oxygen. Oxygen adds to the terminal radical of the growing polystyrene chain by many orders of magnitude faster than the styrene monomer. A relatively stable peroxide radical forms, and further polymerization is effectively prevented. The result is a soft and tacky surface; the interior is not affected, however, since oxygen is not capable of penetrating far into the wet film (see p. 221).

The air inhibition may, of course, be avoided by curing under nitrogen or some other inert gas. This is usually not feasible, however, and two other approaches are used to overcome the

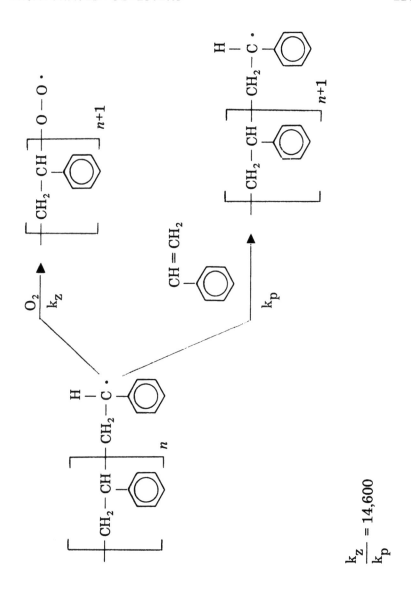

$$\frac{k_z}{k_p} = 14,600$$

problem. In the first of these, a wax-type material is added to the resin solution prior to cure. Once the cross-linking begins the material migrates to the surface where it forms a mechanical barrier, thus preventing oxygen from coming into contact with the styryl radicals. Paraffin is most commonly used for this purpose but various other waxes, or waxlike materials, are also employed. It has been reported that this way of preventing oxygen inhibition gives best effect with polyester resins having low hydroxyl numbers [28].

The formation of the paraffin barrier may be visualized as follows [65]. During the curing process evaporation of styrene takes place from the surface (the polymerization reactions are exothermic), leading to the formation of vortex currents. Paraffin migration to the surface results in supersaturation and precipitation of a wax layer. Part of this layer is transported back into the interior of the film by the vortex current but the remaining precipitate grows over the surface and eventually forms a continuous surface barrier, a few nm thick. After completed cure the paraffin will be enriched at the surface but the bulk of the coating is by no means free from wax, as is shown in Figure 6.17. It can also be seen from this figure that the paraffin concentration is high also at the interface between substrate and coating. This illustrates one of the main problems with the mechanical barrier approach—poor adhesion of the film to the substrate. Another drawback of the wax-type polyesters is the need to remove the wax after curing by some type of polishing.

The other approach used to overcome the air inhibition is to incorporate groups into the system which readily react with oxygen, thus preventing the polystyryl radical-oxygen reaction. Allyl ethers, either incorporated into the polyester resin or employed as a partial replacement for styrene, are widely used for this purpose. The most convenient way of introducing allyl

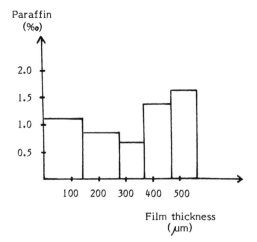

Figure 6.17 Distribution of paraffin in a polyester film. (From Ref. 65.)

ether groups into the polymer is to use polyols partially etherified with allyl alcohol. Typical compounds used are trimethylolpropane diallylether and pentaerythritol monoallyl acetal. Allyl glycidyl ether is another interesting example of the same approach.

$$C_2H_5 \quad CH_2-OCH_2-CH=CH_2$$
$$\underset{C}{\diagdown \diagup}$$
$$\diagup \diagdown$$
$$HOCH_2 \quad CH_2-OCH_2-CH=CH_2$$

Trimethylolpropane diallyl ether

$$HOCH_2 \quad CH_2O$$
$$\diagdown \diagup \quad \diagdown$$
$$\underset{C}{\qquad} \quad CH-CH=CH_2$$
$$\diagup \diagdown \quad \diagup$$
$$HOCH_2 \quad CH_2O$$

Pentaerythritol monoallyl acetal

$$CH_2 - CH - CH_2 - O - CH_2 - CH = CH_2$$
$$\diagdown \diagup$$
$$O$$

Allyl glycidyl ether

Besides allyl ethers, benzyl ethers have also found use as an oxygen-reactive group suitable for incorporation into polyester resins. These two resin types are referred to as air-drying unsaturated polyesters, and they have been in use for almost 30 years. Compared to the paraffin wax approach, the use of air-drying groups incorporated into the polyester gives a better surface finish with higher gloss.

Recent investigations on the mechanism of the curing of allyl and benzyl modified polyesters show that in the inner parts of the film a normal polymerization between fumaric or maleic esters and the reactive diluent is taking place. The air-drying groups show little participation in the polymerization in the bulk of the coatings. At the surface, however, reaction between the reactive diluent and the allyl or benzyl group occurs [34]. The mechanism of the autoxidation seems to involve formation of hydroperoxides in the allylic or benzylic position, as shown in Figure 6.18 for allyl ethers. Decomposition of the hydroperoxide, which is catalyzed by cobalt salts, leads to free radicals (*1, 2,* and *3*), which may copolymerize with the fumarate or maleate unsaturation of the polyester backbone or with the reactive diluent. In this way, oxygen at the surface is consumed through hydroperoxide formation rather than by generation of chain-terminating styryl-peroxide radicals.

Analysis of the gas evolved during curing of allyl ether modified unsaturated polyester resins indicates that the allyl ether group is partially decomposed. Acetaldehyde, acrolein, and isopropanol are formed in considerable amounts. These decomposi-

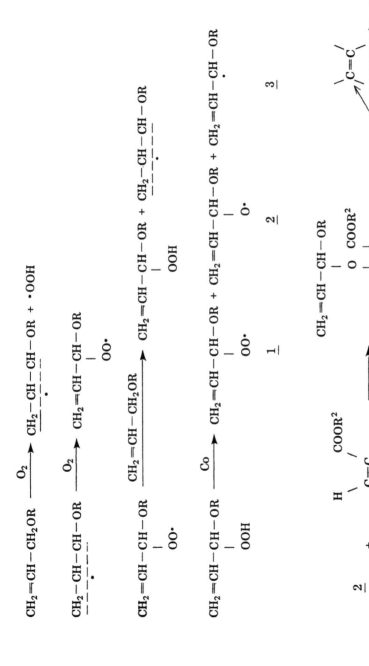

Figure 6.18 Curing of an allyl ether modified unsaturated polyester ("air-drying unsaturated polyester").

225

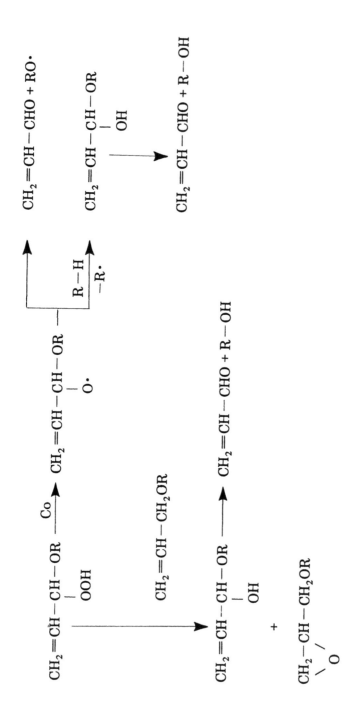

Figure 6.19 Decomposition of allyl ether hydroperoxides.

tion reactions may be regarded as side reactions, competing with the curing reactions of Figure 6.18. The formation of acrolein can be accounted for by different mechanisms, as is shown in Figure 6.19. The reaction pathways leading to the other low molecular products are not as obvious [34].

REFERENCES

1. D. H. Solomon, *The Chemistry of Organic Film Formers*, Krieger, New York, 1977, pp. 46-51,56,143.

2. J. R. Chipault, E. C. Nickell, and W. O. Lundberg, *Off. Dig.*, *23*:740 (1951).

3. E. H. Farmer and D. A. Sutton, *J. Chem. Soc.*, 10 (1946).

4. T. P. Hilditch, *J. Oil Col. Chem. Assoc.*, *30*:1 (1947).

5. N. A. Khan, *Can. J. Chem.*, *37*:1029 (1959).

6. O. S. Privett, C. Nickell, W. E. Tolberg, R. F. Paschke, D. H. Wheeler, and W. O. Lundberg, *J. Amer. Chem. Soc.*, *31*:23 (1954).

7. M. M. Horikx, *J. Appl. Chem.*, *14*:50 (1964).

8. O. S. Privett and C. Nickell, *J. Amer. Oil Chem. Soc.*, *33*:156 (1956).

9. W. Carruthers, *Some Modern Methods of Organic Synthesis*, Cambridge University Press, London, 1971, p. 128.

10. L. Williamson, *J. Appl. Chem.*, *3*:301 (1953).

11. C. E. H. Bawn, *J. Oil Col. Chem. Assoc.*, *40*:1027 (1957).

12. S. Enomoto, H. Takita, S. Nishida, H. Wada, Y. Mukaida, and M. Yanaka, *J. Appl. Polym. Sci.*, *22*:253 (1978).

13. L. A. O'Neill, *Chem. Ind.*, 384 (1954).

14. H. W. Chatfield, *Paint Technol.*, *26*:17 (1962).

15. H. Schulz and H. Wagner, *Angew. Chem.*, *62*:105 (1950).

16. S. Hochberg, *J. Oil Col. Chem. Assoc.*, *48*:1043 (1965).

17. French patent 1,210,192 (1959).

18. French patent 1,257,246 (1961).

19. E. Levine, E. J. Kuzma, and M. T. Nowak, *Mod. Paint Coat.*, *66*:23 (Aug. 1976).

20. K. E. J. Barrett and R. Lambourne, *J. Oil Col. Chem. Assoc.*, *49*:443 (1966).

21. D. B. Larson and W. D. Emmons, *J. Coat. Technol.*, *55*:49 (1983).

22. U.S. Patent 2,414,089 (1947).

23. V. F. Jenkins, A. Mott, and R. J. Wicker, *J. Oil Col. Chem. Assoc.*, *44*:42 (1961).

24. P. H. Seiden, *Glasfaserverstärkte Kunststoffe*, Springer Verlag, Berlin, 1967.

25. H. V. Boenig, in *Encyclopedia of Polymer Science and Technology* (N. M. Bikales, ed.), Vol. 11, Interscience Publ., New York, 1969.

26. H. V. Boenig, *Unsaturated Polyesters: Structure and Properties*, Elsevier, Amsterdam, 1964.

27. N. H. Proctor and J. P. Hughes, *Chemical Hazards of the Workplace*, J. B. Lippincott, Philadelphia, 1978.

28. J. Mleziva, *Farbe Lack*, *86*:689 (1980).

29. P. E. Froehling, *J. Appl. Polym. Sci.*, *27*:3577 (1982).

30. W. D. Cook and O. Delatycki, *J. Polym. Sci.*, *12*:1925 (1974).

31. K. Horie, I. Mita, and H. Kambe, *J. Polym. Sci.*, *8*:2839 (1970).

32. T. R. Cuadrado, J. Borrajo, R. J. J. Williams, and F. M. Clara, *J. Appl. Polym. Sci.*, *28*:485 (1983).

33. A. W. Birley, J. V. Dawkins, D. Kyriacos, and A. Bunn, *Polymer*, *22*:812 (1981).

34. H.-J. Traenckner and H. U. Pohl, *Angew. Makromol. Chem.*, *108*:61 (1982).

35. A. Kastanek, J. Zelenka, and K. Hajek, *J. Appl. Polym. Sci.*, *26*:4117 (1981).

36. M. Vecera, V. Machacek, and J. Mleziva, *Sb. Prednasek, Makrotest, Celostatni Konf.*, *4:th*, *2*:93 (1976); Chemical Abstract, *86*:56015b (1977).

37. B. T. Hayes and R. F. Hunter, *Chem. Ind.*, 559 (1957).

38. A. Fradet and E. Marechal, *Macromol. Chem.*, *183*:319 (1982).

39. T. Okita and S. Oishi, *J. Chem. Soc. Japan, Ind. Chem. Sect.*, *58*:315 (1955).

40. S. K. Gupta and R. T. Thampy, *Macromol. Chem.*, *139*:103 (1970).

41. Z. Ordelt, V. Novak, and B. Kratky, *Collect. Czech. Chem. Comm.*, *33*:405 (1968).

42. S. Knodler, W. Funke, and K. Hamann, *Makromol. Chem.*, *53*:212 (1962).

43. Z. Ordelt, *Makromol. Chem.*, *63*:153 (1963).

44. Z. Klosowska-Wolkowicz and E. Kicko-Walczak, *Plast. Massy*, 53 (1983); Chemical Abstract, *99*:213280x (1983).

45. D. Ya. Filippenko, N. V. Khrenova, and A. Kh. Bulai, *Polimery*, *24*:391 (1979); Chemical Abstract, *92*:181937d (1979).

46. W. E. Douglas and G. Pritchard, *Br. Polym. J.*, *15*:19 (1983).

47. *Polymer Handbook* (J. Brandrup and E. H. Immergut, eds.), Interscience, New York, 1966, p. II-230.

48. J. C. Bevington, M. Johnson, and J. P. Sheen, *Europ. Polym. J.*, *7*:1147 (1971).

49. M. Gordon, B. M. Grieveson, and I. D. McMillan, *J. Polym. Sci.*, *18*:497 (1955).

50. Anon., *Res. Discl.*, *239*:98 (1984).

51. S. A. Bratslavskaya, N. G. Videnina, and K. V. Zapunnaya, *Kompoz. Polim. Mater*, 3 (1980); Chemical Abstract, *95*: 82442u (1981).

52. S. I. Omel'chenko, N. G. Videnina, A. S. Rot, and G. V. Shiryaeva, *Khim. Tekhnol.*, 8 (1981); Chemical Abstract, *95*:220727u (1981).

53. S. S. Jada, *polym. Prepr.*, *22*:2 (1981).

54. L. H. Sperling, *Macromol. Rev.*, *12*:141 (1977).

55. L. Y. Young, in *Copolymerization* (G. E. Ham, ed.), Interscience, New York, 1964, p. 854.

56. V. R. Kamath and R. B. Gallagher, *Plast. Comp.*, *4*:41 (1981).

57. Anon, *Plast. Eng.*, *32*:26 (1976).

58. K. Demmler and J. Schlag, *Farbe Lack*, *77*:224 (1971).

59. L. Horner and K. Scherf, *Justus Liebigs Ann. Chem.*, *573*: 35 (1951) and *574*:202 (1951).

60. W. H. Canty, G. K. Wheeler, and R. R. Myers, *Ind. Eng. Chem.*, *52*:67 (1960).

61. J. P. Cassoni, G. A. Harpell, P. C. Wang, and A. H. Zupa, *Plast. World*, *35*:55 (1977).

62. C. U. Pittman, Jr. and S. S. Jada, *Ind. Eng. Chem. Prod. Res. Dev.*, *21*:281 (1982).

63. S. S. Jada, *Makromol. Chem.*, *183*:1763 (1982).

64. E. W. Lord, R. G. Rice, and E. E. Stahly, *Ind. Eng. Chem. Prod. Res. Dev.*, *10*:391 (1971).

65. K. Demmler, E. Mueller, and M. Schwarz, *DEFAZET-Dtsch. Farben-Z.*, *31*:115 (1977).

64. E. W. Lord, R. G. Rice, and E. E. Stahly, *Ind. Eng. Chem. Prod. Res. Dev.*, *10*:391 (1971).

65. K. Demmler, E. Mueller, and M. Schwarz, *DEFAZET-Dtsch. Farben-Z.*, *31*:115 (1977).

Index